All About Ham Radio

All About Ham Radio

by Harry Helms, AA6FW

A DX/SWL Press Book

HighText

publications inc.

San Diego

Cover design and illustrations: Brian McMurdo, Ventana Studio, Valley Center, CA
Developmental editing: Carol Lewis, HighText Publications
Production services: Greg Calvert, Artifax Corporation, San Diego, CA

ISBN: 1–878707–04–3
Library of Congress catalog number: 92–070472

HighText is a trademark of HighText Publications, Inc.

Contents

Foreword

Just what's so fascinating about ham radio?

You're probably asking that question. Since you've probably already paid for this book (or are at least thinking of buying it), I hate to admit that I don't have a good answer to that question even after over two decades as a ham radio operator!

But clearly something about ham radio is seductive and a lot of fun. Why else would I spend so many hours behind the dials of my ham radio equipment? Why else would I go to such trouble erecting and tuning antennas? Why else would I forget to take a razor or toothbrush along on a trip, but remember to take along my ham radio walkie-talkie—and also remember to fully recharge its batteries the night before???

Part of the fun of ham radio is its unlimited possibilities. You never know who you're going to find yourself talking to when you pick up the microphone of a ham radio set. Ever had the chance to talk to someone living in the south of France, Tahiti, Moscow, rural Argentina, and Montana in the same day? A lot of hams do it all the time and don't think twice about it. Many hams talk around the world over communications satellites built by hams just for other hams. A growing number of hams "talk" to each other through their computers, linked together in extensive on-the-air networks. You can even operate your own ham television station. And a lot of hams just enjoy talking to people in their home town using small, inexpensive walkie-talkie units.

"It offers something for everyone" is a cliche, but it's true when it comes to ham radio. Like experimenting, tinkering, and trying out the latest high-tech gizmo? If so, ham radio will keep you busy and out of trouble for the rest of your life. Feeling competitive? There are numerous on-the-air contests where you try to contact as many different states, countries, geographic zones, U.S. counties, stations in a particular state, etc., as you can. Like to help others? Public service has long been a tradition in ham radio, with hams furnishing communications for public events such as parades, festivals, marathon races, etc. And you might like just talking to a wide spectrum of people from all walks of life. If so, ham radio is definitely for you!

While it's fun, there's a serious side to ham radio. Whenever there's an emergency—such as the twin natural disasters of Hurricane Hugo and the San Francisco earthquake in 1989—ham radio is often the only communications link that still functions. Many hams practice just for such emergencies, and networks of ham radio operators quickly spring into action to facilitate handling of messages into and out of the stricken areas. I listen avidly to such networks, and often find the information I hear on them to be more reliable and timely than that from radio and television newscasts. Sometimes you can follow incredibly dramatic events through ham radio. For example, in January of 1991 ham radio operators throughout North America were able to listen to fellow ham radio operators in Vilnius, Lithuania describe the movement of Soviet troops upon governmental buildings in that city. And a few months later Russian hams inside the besieged Russian parliament building provided vital communications links for Boris Yeltsin and other leaders as they organized and directed opposition to the attempted Soviet coup. By early 1992, hams got to hear old friends in the former USSR proudly using new call letters issued by the newly independent republics.

There are some unusual side benefits of being a ham radio operator. After a couple of years of being a ham, you'll probably become a master of obscure geographical trivia and be able to run the entire "Geography" category on the television program *Jeopardy*. Try locating Cabinda on a map, for example. Or St. Pierre et Miquelon, Ceuta, Jan Mayen Island, and

Tonga. Can't do it? A lot of hams could find them in their sleep. And you'll accumulate (without trying to) a vast array of scientific and technical esoterica. You'll know the layers of the ionosphere, the effects of solar flares, what impedance is, how ducts form in the troposphere, the relationships between voltage and current, and a zillion other scientific things your teachers tried unsuccessfully to teach you in school.

People of all different ages, sizes, colors, and shapes get bit by the ham radio bug. Television newscaster Walter Cronkite, King Hussein of Jordan, rock guitarist Joe Walsh, the late Yuri Gagarin (the first man in space), actor Marlon Brando, former Arizona senator and presidential candidate Barry Goldwater, several astronauts and cosmonauts, and myself don't have much else in common, but we're all ham radio operators.

There's never been a better time to join us on ham radio. The Federal Communications Commission (FCC) has finally authorized a ham radio license that doesn't require you to take a Morse code test. You do have to pass a written exam, but it's on basic operating procedures and other stuff you'll need to know anyway to properly operate a ham radio station. (Elementary school kids pass the test each week, so it's not that hard.) Equipment, priced in today's inflation-ravaged dollarettes, takes a smaller bite out of your budget than ever before—if you can afford a half-way decent stereo system or personal computer, you can buy a complete ham radio station and have change to spare. Today's ham radio equipment is compact, smartly designed, and easy to use, not some weird collection of stuff that looks like it belongs on the Starship *Enterprise*. And the ham radio bands are filled with colorful, interesting people who are eager to meet you.

What more can I say? The bottom line is that ham radio is fun! On the air, my call letters are AA6FW. Look for me when (that's *when*—not *if!*) you have your license. I'd like to get to know you better and hear what's new with you.

Harry L. Helms

A Dozen Unfortunate Myths about Ham Radio

I T SEEMS THAT EVERYBODY'S HEARD about ham radio, but not that many people really understand what it's all about. The mainstream media doesn't help much, since they often get it confused with CB radio or otherwise screw up the facts. Over the years, I've discovered that a lot of people who are interested in ham radio have some really mistaken ideas about ham radio. Sadly, most of these ideas discourage these people from exploring ham radio further. To set the record straight, here is my admittedly subjective list of "Dirty Dozen" myths about ham radio, along with the real scoop:

1. You have to learn Morse code to get a ham license.

False!!! It used to be that way, but in 1991 the Federal Communications Commission (FCC) eliminated the Morse code requirement for the Technician class ham radio license. All you have to do to get a Technician license is pass a written exam. The Technician class license lets you use voice and do a lot of different things, such as operate television, communicate through ham radio satellites, and use computer-to-computer communications (called "packet radio"). If you've been putting off getting a ham radio license because of the Morse code requirement, you've got no reason to hesitate any longer!

2. You have to be an electronics genius to pass the written test for a ham license.

False! The written exams for the two "entry" levels of ham licenses, the Novice and Technician classes, are passed each year by lots of junior high school kids. Instead of obscure electronics theory, the written exams deal with the practical stuff you'll need to know anyway to properly set up and operate your station. I'll admit some already-licensed hams like to boast about how difficult the test was and how hard they had to work to pass it, but they're just exaggerating their accomplishments. (I've passed all the different levels of written exams for ham licenses; they can't be that hard if a dummy like me can pass them!) All tests are the multiple-choice type, and all questions (and answers) are drawn from a large pool of questions which are released to the public by the FCC. If your memory is good enough, you could pass the written exam strictly by memorization!

But I have to let you in on a secret: the written exam requirement is a good filter to eliminate the sort of riff-raff that messed up CB radio for everyone. The fact that you're reading these words means you're probably above average in intelligence, education, and motivation. Ham radio is populated by similar people. Hams come in all ages, races, religions, nationalities, and backgrounds, but the one thing they do have in common is that they tend to be brighter, more interesting people than the rest of the human race. Dullards and slackers can't summon up the energy and effort to pass the written exam.

3. You have to be an electronics genius just to operate a ham radio station.

No! Today's ham radio equipment is no more difficult to use than a VCR or stereo system. I wish I could tell you (just to benefit my ego) that it takes loads of expertise and knob-twisting to get a ham radio station to operate "just right," but in reality all you need to know is how to read and follow instructions in the owner's manual. However, I do admit that as you get more experienced in ham radio, you'll learn a good bit about electronics without really intending to. If you get really carried away, you can be like the thousands of hams who actually build their own equipment using parts they get from Radio Shack and similar places.

4. Ham radio equipment is expensive.

False! You can easily spend more than the cost of a new car on a ham radio station, but you can also get on the air and have fun with new equipment that costs the same as a microwave oven, compact disc player, or other item of consumer electronics. (A new ham radio station costs a lot less than a home video camera!) If you, like a lot of new hams, start out with used equipment, you might be able to get on the air for less than $100!

5. I wouldn't be able to get on the air anyway, because I live in this apartment complex that prohibits outside antennas.

No problem! It's possible to get on the air and communicate over a wide area using simple indoor antennas much like the telescoping "whip" antennas on portable radios. (You can sometimes use even smaller antennas only a few inches long; these are flexible, rubbery antennas known as "rubber duckies.") And you can install ham radio equipment in your car, or carry around a complete ham radio station in your shirt pocket. In all honesty, these "mini antennas" won't work as well as those big outdoor monsters some hams use, but they will be more than adequate for short and medium range work, and some hams have contacted over 100 countries using simple wire antennas located in their attics or taped to their walls. And many ham radio clubs as well as individual hams have installed "repeater" stations which relay signals from lower-powered stations to increase their coverage.

6. I'll interfere with the television and radio reception of my neighbors.

False! You may have heard stories about (or even experienced for yourself) interference to radio and television reception because of "somebody operating their ham radio." In the vast majority of cases, the culprit is actually a citizens band (CB) station using an illegal amplifier to boost their transmitter power. These amplifiers, being illegal, are usually not well constructed and radiate spurious signals which interfere with radio and TV reception. A ham radio station operated in a proper manner seldom causes interference to radio and TV reception. And in those cases where interference is present, it's more often than not due to faulty design of the radio or TV set experiencing the interference. A simple filter on the radio or TV set experiencing reception interference is usually enough to cure the problem. (Many consumer electronics devices are designed without proper filtering circuits to reject interference because manufacturers want to shave a few cents off their per-unit cost.)

7. I'll have to do a lot of soldering and other electrical work to get a ham radio station on the air.

Nope! Most ham radio equipment is like other items of consumer electronics—just plug in the cables and wires to the proper places, and away you go! If you know how to use screwdrivers, pliers, and other simple hand tools, you already have 90% of the mechanical skills required of a ham radio operator. The other 10%, mainly the ability to make a good electrical connection using a soldering iron, is something you can pick up with an afternoon of practice.

8. I'll have to keep a lot of records for the FCC or other government agencies.

False! A few years ago, the FCC required hams to keep a detailed logbook of their operations. That's been eliminated, and now all you basically have to do is to keep your license properly renewed (it's good for ten years at a time) and let the FCC know your current address. And no matter what stupid stories you've heard, the CIA and FBI don't keep tabs on hams who talk to other hams in foreign countries. Nobody's been denied a security clearance for being a ham radio operator.

9. I only speak English, so no one overseas would be able to understand me.

Wrong! English is the universal language of ham radio. When hams in Japan talk to those in Russia, they use English. I've talked to hams in Nigeria, Indonesia, Argentina, Portugal, France, Germany, Denmark, Finland, Greece, Tahiti, and Surinam—and in each case, English was the language of our conversation. On the other hand, if you want to practice a foreign language with someone who speaks it as their mother tongue, ham radio offers excellent opportunities.

10. It takes a lot of study time to get a ham radio license, and you have to travel to an FCC office a couple of hundred miles away to take the test, which they only give a couple of times a year.

Doubly false! How long it takes to learn something varies from individual to individual, but I'd bet that the typical reader of this book would need only about a month—*maximum*—of study to pass the written exam for a Technician class license. (I'm figuring that at a rate of an average of one hour of study per day.) Now admittedly this amount of study won't make you a radio expert or let you understand the ins and outs of ham radio equipment, but it *will* let you pass the license exam. Don't try to (or feel you have to) learn everything about ham radio before you get your license; like driving and marriage, there's a lot you can learn only after you have that license. And license exams are conducted on a volunteer basis by teams of already-licensed hams. Instead of some impersonal government office and employees, exams are conducted on weekends or at nights in friendly surroundings by people who really want to see you pass. (Don't get me wrong; there are strict safeguards to prevent cheating, and you pass—or fail— the exam on your own. But it's like a sports event where you're the home team—people will be rooting for you!) In most areas of the country, there are several exam sessions each month within convenient driving distance.

11. Ham radio is just like CB radio.

No way! Confusing ham radio with CB radio is like confusing China with Japan. Both are countries in Asia with written languages rich in ideograms, but otherwise they're as different as night and day. CB and ham radio both let people communicate with each other by radio, but that's about all they have in common.

CB radio is just what its name implies—"citizens band" radio. It's open to anyone and everyone who can plunk down the money for a CB radio set and antenna. No test or license is required, and the rules are few. (And, to be honest, those rules are *very* seldom enforced.) I'm not putting CB radio down, since I happen to think that average citizens do have the right to use some parts of the radio spectrum without too many restrictions or qualifications. (In the United States, radio frequencies by law belong to the people of the United States; broadcasters, cellular telephones, and ham radio operators all use radio frequencies "loaned" to them by the people of the United States.) But since CB radio is open to everybody, without qualification, it means the capabilities of CB radio have to be greatly restricted to make sure there's room for everyone and to keep down interference (although the sheer number of CB stations means interference is often heavy on all 40 CB channels).

In contrast, entry into ham radio requires passing a test. This means there are fewer hams than CB operators, that hams are better qualified to use high-powered transmitting equipment, and that the discipline among hams is much higher than CB operators. This is reflected in the privileges available to ham and CB operators. For example, CB radio sets are restricted to only 4 watts of transmitter power while hams with a Technician class or higher license can use up to 1500 watts of transmitter power. CB radio operators are restricted to just using voice. Ham radio operators can use voice, Morse code, television, satellites, radio-teletype, packet (computer-to-computer) radio, and other methods of communications. Hams can set up "repeater" relay stations to improve their communications range; CB operators cannot. CB operators are restricted to communications over a maximum range of 150 miles, while hams can communicate all over the world. CB operators are restricted to 40 frequencies (as fixed channels) for communications, while ham operators have hundreds of thousands of frequencies they can use.

Ham radio is more than different from CB radio; it's broader, more versatile, more interesting, and more fun. Let me put it this way: CB radio is a bicycle; ham radio is a Porsche Turbo.

12. I won't fit in.

There's no such thing as a "typical" ham radio operator. They come in all shapes, sizes, ages, colors, sexes, religions, backgrounds, and nationalities. And none of that matters. All that you'll be judged on is how well you can operate your equipment. One of the things that still marvels me about ham radio is how it transcends traditional barriers to human communication and understanding, and allows people who at first glance would seem to have little in common to meet each other.

Okay, so we've exploded some myths about ham radio. Let's find out about the realities.

CHAPTER ONE

Hams and Their Radios

TO BE HONEST, "ham radio" is the wrong name for the subject matter of this book. Under United States law and international treaty, the correct name is *amateur radio* and the *amateur radio service*. However, I'll call the amateur radio service "ham radio" in this book. The term "amateur radio" has always struck me as too formal and a little bit pretentious, while "ham radio" seems friendlier and more fun. And while it is a "service" of sorts to the public, the reason people participate in ham radio is because it's a hobby that's fun as well as being useful. That's why—with all due respect to the people at the Federal Communications Commission who refer to "amateur radio" as a "service"—this book is about the *hobby of ham radio*. In this chapter, we'll take a look at the big picture of ham radio and try to get a feeling for what it's like being a ham radio operator.

Ham Radio Defined (Sort Of. . .)

Okay, so we'll call it "ham radio" instead of "amateur radio." But what's ham radio? It's defined in current FCC regulations this way:

> • A radio communication service for the purpose of self-training, intercommunication, and technical investigations carried out by amateurs, that is, duly authorized persons interested in radio technique solely with a personal aim and without pecuniary interest.

Buried in that governmentese are a couple of key points. One is *intercommunication*. Except for bona fide emergencies, ham stations can only contact other ham stations. You can't contact the U.S. Navy or Air Force if you have a ham radio station. You can't talk to your cousin on his CB radio using your ham station and license. And you can't use your ham station to broadcast to your neighborhood. You're basically restricted to communicating with other ham stations. You can transmit "one-way" signals for such purposes as remote control, testing of station equipment, and as a radio "beacon" so other hams can determine reception conditions, but ham radio isn't a place where you can play disk jockey or say "hi!" to radio officers on ocean liners. You have to communicate with other hams.

(This is as good a place as any to mention that music is strictly *verboten* over any ham station. You can transmit audio tones to test and experiment with transmitting equipment, but don't try to sneak some music past the FCC under the guise of "testing." They've heard that rationale before, and they don't buy it!)

Another important thing in the FCC's definition is *without pecuniary interest*. That means, in general, you can't operate a ham radio station for hire or for the commercial benefit of another party. In other words, if you're one of a group of ham operators providing communications for a public event (such as a parade), you can't accept any money for your services. Nor can you use ham radio to keep in touch with the restaurant if you deliver pizzas for a living. (Nor can you use ham radio to order a pizza from that restaurant if you're a customer, since that would "commercially benefit" the restaurant!) These regulations concerning pecuniary interest are currently under review by the FCC, and it's possible that they will be loosened to some degree in the near future. But—for now at least—keep your business affairs and ham radio separate.

What Do Hams Do?

So ham radio is a hobby where you communicate with other hams and don't get paid for doing so. But what do you actually *do*? You might think the answer is simply "talk on their radios." That's not exactly the right answer, but it's not entirely wrong either. "Talking on their radios" is an important part of ham radio, but it means a lot of different things to different hams. And many ham activities have nothing to do with "talking." Let's take a look at some of the things you can do with a ham license.

Ragchewing

"Ragchewing" is ham lingo for ordinary conversation, whether done by voice or some other method (such as Morse code or radioteletype, a method of sending printed text by radio). If you think this is the most popular activity in ham radio, you're right!

What do hams talk about? The same things everybody else talks about—the weather, sports, people, politics—you name it. Since hams live all over the world and come from every background imaginable, you can have some memorable conversations. One I had was with a ham on the "Big Island" of Hawaii whose home was located less than a hundred yards from an active lava flow from the Kilauea volcano. I was fascinated by his tales of daily life with a lava flow as a next door neighbor, and his calm acceptance of the possibility that a slight change in the direction of the flow could destroy his home within days. Another ham in Australia gave me a complete description of all the venomous snakes found on his farm and their nocturnal habits; I understood why he spent his evenings safely indoors talking on his ham radio equipment!

Chance remarks have been the springboard for some great conversations. Once I made contact with a ham in Texas who said he was a retired petroleum exploration geologist. I replied he must have traveled widely, and that offhand remark was rewarded with an lengthy, engrossing conversation about the places he had visited in the 1950s—namely, every inhabited continent. He saw the world before it became saturated with franchised hotels and satellite dishes, and it made me realize how just recently much of the world had existed in an essentially seventeenth-century state. Another time I was in contact with a Japanese ham, and mentioned I was still sore from a recent mountain climbing trip. It so happened that the Japanese ham was also a climber, and I listened spellbound as he described his climb of Mt. Fujiyama.

There are so many other memorable conversations. . . . a sound technician at the Universal Amphitheater in Los Angeles with his juicy stories about rock bands he's known, the New York cable television producer, the American retiree living in Tahiti. . . . and—to be honest—there have also been a lot of forgettable, mundane ones. But whatever interests you or your background, there are a lot of hams out there you have something in common with and would enjoy talking to.

DXing

"DXing" is attempting to contact other hams in as many different countries (or other rarely contacted places) as possible and exchanging QSL cards with those hams. (What's a "QSL card"? See "Strange Ham Lingo" in this chapter for the answer.) It's very different from ragchewing with foreign hams because very little actual conversation takes place during DXing. A typical DXing contact (called a "QSO") goes something like this:

> "This is V31AA. QR–Zed?"
> "Alpha alpha six foxtrot whiskey."
> "Foxtrot Whiskey. Okay, you're five by nine. Name is
>   Gordon, QTH is Belize City. QSL?"

"QSL, Gordon. Name is Harry, you're also five by nine in southern California. Thanks for the new one, will QSL via the bureau. 73 from AA6FW."

"Okay, Harry, good luck. Victor 31 alpha alpha, QR–Zed?"

Believe it or not, many hams (including me) are thrilled by contacts like this.

In the previous section, I told of the lengthy, interesting conversations I've had with hams in foreign countries. Why didn't I try to engage V31AA in more normal conversation? Surely there must be something going on in Belize City I'd like to know about. The answer is I would've liked to know more about Gordon and what's going on down in Belize, but the unwritten "rules of DXing" call for keeping contacts as short as possible so other hams will have a chance to also contact V31AA. Is Gordon offended that no one wants to take the time to talk with him at length? No, because Gordon is obviously also a DXer—a ham who plays the "DX game"—and knows the rules; he's out to help as many people as he can make a contact with Belize and get his QSL card. The key to whether a station is DX or not doesn't involve distance; it's about rarity instead. Japan and Australia are both further away than Belize, yet Belize is DX and those two countries aren't. Why? Because Japan and Australia have thousands upon thousands of ham operators using modern equipment and antennas. Most days it's easy to hear and contact dozens of stations from those two countries. By contrast, nations like Belize have very few ham operators (sometimes in the single digits), their equipment is often not the best, and it can be weeks, months, or even years before you have the chance to contact a station in those countries.

DX doesn't always mean contacts with stations on the other side of the world. The different frequency bands used by hams have different normal ranges over which communications can be conducted, and anything beyond the normal communications range for a particular band is considered DX.

For example, there are some ham bands over which the normal maximum communications distance is about 150 miles. Under certain conditions which happen rarely, it's possible to communicate on those bands over distances of hundreds or even thousands of miles. All contacts made during those conditions are also considered "DX."

DXers have a special definition of "country." Besides the usual, everyday meaning, DXers also consider certain political units (such as Puerto Rico and British territories like Bermuda), territory isolated from the main country mass (such as Hawaii and Alaska), and political and geographic oddities (like the United Nations building in New York and the U.S. Navy base at Guantanamo Bay, Cuba) to be separate, distinct "countries." The result is that there are over 300 "radio countries" for hams to contact.

DXers collect countries and the QSL cards from them the same way other people collect coins, stamps, baseball cards, or Rin Tin Tin memorabilia. A very popular award sought by many hams is the "DX Century Club (DXCC)," sponsored by the American Radio Relay League. This award is given to hams who have contacted and received QSL cards from 100 different radio countries. The ARRL also issues endorsements for the DXCC award for additional countries contacted above the initial 100 all the way to a "Honor Roll" for those who have contacted all or almost all available countries.

If you don't already have a ham license, DXing might not seem that much fun and a pretty irrational activity. Why would anyone go to all that trouble and expense just for a quick "hello/goodbye" conversation and the chance to swap a card saying the contact took place? Well, I can't explain the allure of it here. DXing is like Zen—it can't be described; you have to "get it" for yourself. But just imagine you hear a weak, distant signal from somewhere you've never heard before. . . . and there's a crowd of other stations trying to contact that

4U1ITU

INTERNATIONAL AMATEUR
RADIO CLUB
P.O. Box 6 - CH - 1211 GENÈVE 20
Switzerland

Dear OM Harry!

TO	DATE	TIME UTC	BAND MHz	MODE 2X	RST	OPERATOR
AA6FW	*9-12-89*	*1628*	*28*	*SSB*	*5.9*	*Felix* *JL8OBC*

Amateur station 4U1ITU is located at ITU headquarters in Geneva. The International Telecommunication Union is the specialized United Nations agency for telecommunications. It counts 166 Members.

6TH WORLD
TELECOMMUNICATION EXHIBITION
GENEVA 8-15 OCTOBER 1991
Organized by International Telecommunication Union

FIGURE 1-1: Believe it or not, this QSL card confirms a contact made with a ham radio "country"—the International Telecommunications Union headquarters building in Geneva, Switzerland! The United Nations building in New York City is also considered as a separate country. Why does this 4U1ITU card count as an independent country instead of Switzerland—and why aren't other notable structures (like the Empire State Building) separate radio countries? Well, it's a long story, but it boils down to this—sometimes hams do things in strange ways.

station. . . . you give a call. . . . and suddenly you hear your own call letters being repeated by that distant station as he or she answers you. . . . yes, if you can imagine what that moment feels like, that's what DXing is all about!

Certificate Chasing

There are other awards in addition to DXCC sponsored by various amateur radio societies and clubs or magazines, and some hams spend all their time trying to earn them. These awards are normally certificates, but some (especially the more difficult to earn ones) are handsome plaques.

Popular awards include the Worked All States (WAS) and Worked All Continents (WAC) certificates, which require contacting and exchanging QSL cards with hams in all states and all permanently inhabited continents, respectively. CQ Magazine sponsors the Worked All Zones (WAZ) award, in which the world is divided into 40 geographic zones. This award requires hams to literally contact "every corner of the world" and is quite popular with DXers. Other awards are available for contacting U.S. counties, Swiss cantons, Russian oblasts (oblasts are similar to U.S. counties), or even a certain number of different station call sign prefixes. If some ham activity involves the accumulation of a certain number of QSL cards and is at least mildly difficult to do, odds are there's a certificate issued somewhere by someone for doing it!

Tinkering

Some hams spend most of their hobby time working on their equipment and antennas. They make some adjustments or modifications, get on the air just long enough to test the results, and then immediately start tearing into things again. Other hams enjoy restoring and using old vacuum tube ham radio equipment. These hams are like the home handyman who's miserable when there's nothing to fix. In the early days of ham radio, all hams were tinkerers because there was no commercially available equipment, and if you didn't "roll your own" you weren't on the air. Today, putting a ham station together is as easy as connecting a VCR to your television. (Maybe even easier, come to think about it. . . .) But tinkering can get in your blood. There's a real thrill being able to say "it works, and I did it myself!" I'm not immune. I'm always changing or rearranging items of equipment in my ham station, trying out new antennas, and poking around inside my gear to repair it or improve its performance. My repairs and "improvements" aren't always as successful as I'd like, but so what the heck? I had fun trying!

Strange Ham Lingo (Part One of a Continuing Series). . . .

Like all specialized hobbies and activities, ham radio has a nomenclature that seems impenetrable at first glance. Many are three-letter pseudowords beginning with the letter "Q." These are taken from international radiotele-graph abbreviations, and are a holdover from the early days of ham radio when all communications used Morse code. The only way to master "hamspeak" is to take it in small doses. Throughout this book, there'll be more "strange ham lingo" courses to help you understand this new language; with practice (and a little luck!) you'll soon sound like an old timer!

ARRL: acronym for the American Radio Relay League, the national association of ham radio operators. Often referred to simply as "the League."

bureau: a clearinghouse for QSL cards sent to and received from overseas hams operated by a national ham radio organization to help save individual hams postage costs.

CQ: this is a call sent by ham stations indicating they're willing to answer anyone who replies. It's used to start an on-the-air conversation, as in "CQ, CQ, CQ, calling CQ. This is AA6FW calling CQ and listening." When I say that, I'm inviting anyone who hears me to reply. "CQ" is also the title of a popular ham radio magazine which sponsors operating awards and on-the-air contests.

DX: ham radio stations located far away or otherwise difficult to contact, and the sport of trying to contact as many of these stations in different countries, states, Swiss cantons, Japanese prefectures, etc., as possible. Any real ham with the right instincts goes through a period where this is his or her major operating activity. (At least, that's my opinion.)

eyeball: a face-to-face meeting between hams, as in "I eyeballed him at the hamfest" or "we had an eyeball QSO last week." (See "QSO" and "hamfest.")

final: the last transmission by a station during a contact. "I've got to go now, so this will be my final. . . . "

hamfest: a gathering where hams sell and swap equipment, hear speakers, attend forums, meet each other in person, and eat mediocre, fattening food.

handle: a name. "Good to meet you, my handle's Harry."

landline: the telephone or a telephone call. "I'd rather not discuss that on the air; better give me a landline instead."

mobile: a ham station installed in a car, boat, or other vehicle where it can be operated while in motion. Often pronounced "moe–bile" (the *i* is long), as in "I'll be operating moe-bile Saturday." (Often frowned upon by people who understand correct English pronunciation.)

net: short for "network," a group of ham stations that meet on a certain frequency at a certain time. There are nets that meet to handle traffic for third parties, for working DX, for emergency preparedness, or just to let old friends keep in touch with each other.

out: said at the end of a transmission to indicate it is your last transmission and that no reply is expected from the other station.

over: said at the end of a transmission to let the other station know it's their turn to transmit.

patch: a device which connects a ham radio station to the public telephone network.

personal: see "handle." "The personal here is Harry." (Often frowned upon by long-time hams.)

phone: short for "radiotelephone," or voice operation.

QRZ: a call inviting any station listening to answer, usually pronounced "Q–R–Zed" (the British pronunciation) and asked as a question. QRZ differs from CQ in that it is normally used to solicit the next contact in a series of contacts or if a station is not sure who is calling it, as in "QRZ, could you please repeat your call sign?"

QSL: a card or other piece of paper sent to one ham by another ham to confirm that a contact did indeed take place between their stations. Each letter is pronounced individually, as in "Q–S–L." (See "wallpaper" below.) QSLs are used to earn awards for contacting different states, countries, etc.

QSO: this is pronounced "Q–so," and means a contact or conversation between hams.

QST: a general call preceding a transmission addressed to all hams, with no reply expected. It's also the title of the ARRL's monthly magazine sent to its membership.

QTH: the location of a ham radio station.

repeater: a ham radio station which receives and automatically retransmits the signals of another ham radio station, usually to extend the range of hand-held "walkie-talkie" units or mobile stations.

rig: any and all items of equipment in a ham radio station.

shack: for some weird reason, hams refer to the area where their station is located as the "shack." Often, the "shack" is the most comfortable, most luxuriously furnished part of the house or apartment!

sign with: to end a contact with another station. "I have to go, so let me sign with you. . . . "

ticket: a ham radio license. "I got my ticket back in 1967."

traffic: messages from or to third parties exchanged by two ham radio stations. "I heard the communications from the flood area; it was mostly Red Cross traffic."

wallpaper: certificates, QSL cards, or other paper items which could be displayed on the wall of the shack. (See "shack" above.)

work: to contact a station. "Yeah, conditions were really good last night. I worked Indonesia!"

73: the old radiotelegraph code for "best regards," often sent by hams at the end of a contact. ". . . . well, let me send my 73 to you. . . . " Often intensified by hams as "best 73" (best best regards), "73s" (best regardses), and the legendary "very best 73s" (very best best regardses). There's also a ham radio magazine titled *73*.

88: the old radiotelegraph code for "love and kisses." Should only be used between hams of the opposite sex unless one desires to upset and/or titillate anyone who might be listening, not to mention the ham one is in contact with.

More to come in various forms throughout this book. . . .

Contesting

Ham radio organizations and magazines sponsor several on-the-air contests for hams. These involve trying to contact as many countries, states, zones, or other "radio entities" as possible within a certain period of time, usually one or two days. Trophies and certificates are awarded to the highest scoring hams in various categories (such as transmitter power) or in different areas.

Hams who are deeply into contests are known as "contesters." These hams usually have powerful, well-engineered stations with efficient antenna systems and put booming signals into most areas of the world. (Not surprisingly, a lot of DXers have similar stations.) Many contesters only operate during contests, and they play to win; you can sometimes hear them making contacts at the rate of four to five a minute! Top contesters are *very* serious about winning; the competition between top contesters is often furious, and an on-the-air contest can seem a lot more like a war than a hobby. However, it's all in fun, and you can often hear the "major leaguers" discussing their scores with each other a few minutes after the end of a contest.

Both contesters and DXers participate in contests. Many DX stations make it a point to get on the air for major contests, and the rapid pace of a contest makes it possible to work numerous DX stations in quick order.

Public Service

The FCC says ham radio is a service. I say it's a hobby. I guess we're both right.

The public services rendered by hams run from the dramatic (being the only functioning communications link in times of a natural disaster) to the everyday (manning checkpoints for a marathon race and relaying results to scorers). These free communications services are worth millions of dollars to taxpayers each year and provide an essential "glue" for many activities. You've probably been the beneficiary of some ham communications services without realizing it if you've attended a major non-profit public event, such as a parade or bike race, recently.

The ability of hams to respond quickly to emergencies and take over essential communications is no accident. Many hams participate in training activities, such as simulated emergency drills, to hone their abilities to perform in actual situations. Other hams meet on the air in "nets" (short for networks) where they learn how to accept, forward, and process messages for third parties. While these hams are organized and disciplined, they're not bureaucratic; hams can often respond faster and with greater flexibility than official agencies because of the lack of red tape. Hams don't get any payment or other compensation for these activities, but a real ham doesn't need much reward other than a chance to play with his or her radio!

In areas of the United States where tornadoes are common, hams form groups of "weather spotters." When conditions are right for the formation of tornadoes or other violent storms, hams deploy to various locations with good visibility, such as the upper floors of tall buildings, where they can look for signs of tornado formation or other storm activity. These hams are in communication with weather service and local emergency officials, and often provide the first warnings of threatening weather. If a tornado is actually sighted, hams provide crucial information about its location and movement.

Hams frequently run phone patches (see "Strange Ham Lingo") between U.S. military personnel overseas and their stateside families. These are often the only way for military personnel to speak to their families, particularly if they are stationed in isolated areas or aboard ships.

One interesting aspect of public service is how DXers and contesters often play major roles in emergency communications, particularly international emergencies, even though they seldom participate in training activities. However, DXers and contesters are experienced in communicating in pressurized, high interference situations, and their stations deliver the strong, reliable signals essential for emergency communications.

Experimenting

Most hams can't leave well enough alone. They're always trying out new communications techniques and systems, poking around inside their radios, or stringing up some weird new antenna. Often these innovations work a lot better than the stuff coming out of professional research and development labs.

In fact, hams have pioneered many technologies. The most recent example is packet radio, a method by which computers can be linked together by radio. Packet radio allows automatic relay of messages between computers and even asks for retransmission of messages that are incorrectly received. Within the past decade, hams developed interface units for connection between ham radio stations and personal computers and wrote software to control packet radio operations. The U.S. military is now one of the big users of packet radio. In the 1960s and 1970s, hams showed that communications through so-called "low Earth orbit" satellites using simple antennas and equipment was possible. That's now one of the hottest areas of commercial telecommunications activity. Other hams are currently doing remarkable work in such areas as communication by bouncing signals off the moon (!!!) and bending of radio signals in the tropospheric layer of the Earth's atmosphere.

To be honest, most of these innovations were developed by hams who also happened to be professional design engineers; you don't have to be smart enough to do something like that to get a ham license! However, ordinary hams are willing to try techniques and circuits that com-

FIGURE 1-2: An essential tool for the ham radio operator involved in public service activities is a pocket-sized walkie-talkie like this Kenwood unit for the 2-meter band.

mercial firms judge too risky or speculative. Hams also don't have to seek authorization from corporate committees before trying new techniques and systems, meaning hams are more flexible in "field testing" and improving new technologies. The result is that ham radio is where tomorrow's communications technologies are perfected today.

Dang Near Anything Having to Do with Staying in Touch by Radio

If none of the activities we've mentioned so far appeal to you, go make up your own.

Ham radio can be a reliable, long range, and low cost communications medium supplementing your other hobby activities and interests. For example, are you an outdoors enthusiast enjoying activities such as backpacking or mountaineering? You've probably wished for some way to stay in touch with other members of your party—or summon help in an emergency—without the limited range and heavy interference that plagues CB radio. Ham radio is the answer. Are you a computer hobbyist who would like to be able to check into your home computer or a bulletin board system (BBS) without having to go through the expense of a cellular telephone call? Again, ham radio can do it. Maybe your buddies have seats on the other side of the stadium, but you'd like to stay in touch with them during the game. A couple of ham radio walkie-talkies small enough to fit in your shirt pocket will let you set up a halftime rendezvous near a concession stand. Remote control of devices and objects? Of course! Staying in touch with members of your family around town? Sure! In fact, anything that you can do with radio, just as long as it's not broadcasting to the public or for business purposes, can be done with ham radio. And it can usually be done better and cheaper to boot.

The Basis and Purpose of Ham Radio

In addition to defining ham radio, the FCC has also included in its rules five principles forming the basis and purpose of ham radio. These principles are reflected in the FCC's rules for ham radio, and you're likely to be asked a question based on the five principles somewhere on the written exam for a ham radio license, so you might as well memorize them:

(a) Recognition and enhancement of the value of the amateur service to the public as a voluntary, noncommercial, communication service, particularly with respect to providing emergency communications.

(b) Continuation and extension of the amateur's proven ability to contribute to the advancement of the radio art.

(c) Encouragement and improvement of the amateur service through rules which provide for advancing skills in both the communication and technical phases of the art.

(d) Expansion of the existing reservoir within the amateur service of trained operators, technicians, and electronics experts.

(e) Continuation and extension of the amateur's unique ability to enhance international good will.

Notice how the activities noted in the preceding section largely support these five principles? It's not a case of the FCC mandating this, but rather a happy example of expressed aims coinciding with real-world outcomes!

Kilohertz, Megahertz, and Meter Bands

Like every specialized activity, ham radio has its own set of special expressions and terms not found elsewhere. Some of these will be explained in this and other chapters under "Strange Ham Lingo." In other cases, we'll explain them in the text and try to introduce you to them as smoothly as possible.

Before going much further, we need to discuss kilohertz (kHz), megahertz (MHz), and meter bands. These terms are used to measure and describe operating frequencies for ham stations, and we have to use them when talking about the frequency bands hams can use.

We'll discuss the technical details of kHz and MHz later, but for now just remember that *hertz* (Hz) is a measure of how many cycles of a radio wave are transmitted in one second (i.e., its frequency). A kilohertz is 1000 Hz. A megahertz is 1000 kHz. In other words,

$$1,000,000 \text{ Hz} = 1000 \text{ kHz} = 1 \text{ MHz}$$

so the frequencies of 7150 kHz and 7.15 MHz mean the same thing.

The cycles of a radio wave are sent one after the other. The "distance" between the identical points on two consecutive radio waves is known as *wavelength*. Even though radio waves are invisible, there's a distance between the cycles—i.e., the individual "waves"—of a radio signal, just as there's a distance between ocean waves. (In fact, that's a good way to visualize radio waves: as waves of electromagnetic energy arriving at an antenna much like ocean waves arrive at a beach, one after another.) The distance between the peaks of cycles is measured in meters. If we say a radio station "is on 80 meters," we mean that its operating frequency is such that each cycle of its signal is separated by 80 meters.

A higher frequency (kHz or MHz) means more cycles of a radio wave are transmitted in a second. This also means the wavelength between cycles is less—as the frequency in kHz or MHz rises, the wavelength decreases. The relationship between frequency and wavelength can be expressed by the following formula:

$$\text{wavelength} = 300/\text{frequency in MHz}$$

Let's go back to our previous example frequency of 7.15 MHz. By plugging it into this formula, we discover that the wavelength of 7.15 MHz (a.k.a. 7150 kHz) is 42 meters. Looking at it from the other side of the fence, the frequency of 42 meters is 7.15 MHz, which can also be expressed as 7150 kHz.

"Wavelength" is a holdover from the early days of radio, and today survives as a handy shorthand to refer to the ham bands. For example, there's a ham band from 50000 to 54000 kHz, and it's known among hams as the 6-meter band (often simply called "6"). Which of the following statements would you rather say?

"Listen for me tonight on the 50000 to 54000 kHz band."

"Listen for me tonight on 6."

Yeah, me too. That's why hams still refer to their operating bands in meters. Table 1-1 gives the most popular ham radio bands in kHz and MHz along with the meter band equivalents hams refer to them as. (Notice that at some higher frequencies, with their shorter wavelengths, the equivalents are in centimeters instead of meters.)

TABLE 1-1

The Major Ham Radio Bands

Frequency Range	Meter Band
1800 to 2000 kHz	160 meters
3500 to 4000 kHz	80 meters
7000 to 7300 kHz	40 meters
10100 to 10150 kHz	30 meters
14000 to 14350 kHz	20 meters
18068 to 18168 kHz	17 meters
21000 to 21450 kHz	15 meters
24890 to 24990 kHz	12 meters
28000 to 29700 kHz	10 meters
50 to 54 MHz	6 meters
144 to 148 MHz	2 meters
222 to 225 MHz	1.25 meters
420 to 450 MHz	70 centimeters
902 to 928 MHz	33 centimeters
1240 to 1300 MHz	23 centimeters

There are some terms and abbreviations commonly used to refer to broad frequency ranges. All ham bands from 160 to 10 meters are known as the *high frequency* (HF) bands. The 6, 2, and 1.25 meter bands are known as the *very high frequency* (VHF) range. And all ham bands about 70 centimeters in frequency are known as *ultra high frequency* (UHF) bands.

What Equipment Do I Need? (And How Much Does It Cost?)

You do need special equipment for ham radio, but it's nowhere near as elaborate, complicated, or expensive as you might have heard. In fact, you might already have a home video or audio system that's more expensive and complex than many ham radio stations.

Obviously, you'll need equipment to send and receive radio signals. A device that sends radio signals is a *transmitter* while a device that receives signals is a *receiver* (how appropriate!). These are often combined into a single device known as a *transceiver*. Transceivers often need a separate *power supply* unit if they are to be operated off the AC current in your home. Another item you'll need is an *antenna* to allow you to send and transmit signals. Add a microphone or interface device for your personal computer and you're in business!

Ham stations come in all sizes and configurations. The equipment can range from a small hand-held "walkie-talkie" to enormous arrangements of hardware filling an entire room. However, you can fit a transceiver capable of international communications in a space less than that required for a microwave oven; some (such as the transceiver I use) are only as big as a typical shoebox.

FIGURE 1-3: Just add an antenna and you're ready to talk all over the world with this station from Kenwood! At the left is a microphone intended for use at your home station (smaller hand-held "mikes" could also be used). Next to it is a speaker, although all current transceivers come with a built-in speaker. However, an external speaker usually sounds better. The heart of the station is the TS-450S transceiver in the middle. This transceiver lets you operate from 160 to 10 meters using voice, Morse code, and different teleprinter modes. At the left is a power supply for the TS-450S. This entire station occupies only a little more space than a typical compact disc player and stereo receiver!

While finding space for a transceiver is usually no problem, you might have problems with the antenna. This is particularly true if you live in a condo or other housing development where installing outside antennas is prohibited, difficult, or impractical, or want to operate on the 160 to 10 meter bands. However, it is possible to install remarkably effective antennas in smaller areas or indoors if the building does not have a metal frame. Even small indoor antennas can do a fine job if you operate through *repeaters*, which are relay stations operated by hams on 10 meters and higher frequency bands to extend the coverage of lower-powered and mobile stations. The necessary physical size of an antenna decreases with frequency; a full-size, "no compromise" antenna for the two meter band is much smaller than an equivalent antenna for 80 meters.

Ham radio is not always an expensive hobby, although most hams have a minimum of at least $1000 invested in their stations. Most of this investment takes place over several years, however. It's possible to get started in ham radio with new equipment for less than $500, and your expenditure can be under $100 if you buy used equipment for your first station. It isn't a bad idea to start off with used equipment; most works as well as new, more modern gear and you can find out if ham radio is "your thing" before spending serious money on new station equipment.

Besides a transmitter/receiver or transceiver and antenna, you'll likely flesh out your station with accessories such as headphones, microphones, cables, and related goodies. And there's one other item you'll need—a license!

All about Licenses

To get a ham radio license, you have to pass an examination conducted under FCC rules. You might have heard that this is some sort of major hurdle. Relax. It's not.

Since you don't need a license to operate a CB radio or a cellular telephone, you might be wondering why you need one to operate a ham radio station. The big reason is because the power and capabilities of ham radio gear are so great that it's possible for an incompetent user of such equipment to cause real problems for others, including disrupting radio and television reception and interfering with vital aeronautical, police, and fire communications. Can you imagine what our highways would be like if no driver licenses were required? That's how the airwaves would be if no ham license was needed—an overgrown version of the CB radio channels and the bedlam on them.

A ham radio license is actually two parts. One is the *operator* license. The operator license allows certain operating privileges, such as the frequencies that can be used, depending upon the level of examination passed. You can operate any amateur station, whether your own or not, only up to the level of operating privileges authorized by your operator's license. The *station* license is the call letters you use on the air. A station license is issued for a fixed location shown on your ham license, and you also use those call letters when operating mobile or portable station equipment away from the fixed location. Both the operator and station license are contained on a single license form issued by the FCC. Licenses are normally valid for ten years and can be renewed without any further examination or operating activity requirements.

There are currently five classes of ham radio licenses issued by the FCC, with each having its own examination requirements and offering more operating privileges as you climb the licensing "ladder." Each step up the ladder requires a more difficult written exam, and sometimes a Morse code test is required. The current license classes are these:

Novice: This is one of the two "entry level" licenses. It requires a Morse code test at a rate of five words-per-minute (WPM) and a 30 question examination, of which 22 questions must be answered correctly. The Novice class license lets you use Morse code from 3675 to 3725, 7100 to 7150, 21100 to 21200, and 28100 to 28300 kHz. Novices may also use radioteletype and packet radio in the 28100 to 28300 kHz range. Novices are also allowed voice (or *phone*) privileges in the 28300 to 28500 kHz, 222.1 to 223.91 MHz, and 1270 to 1295 MHz ranges. In the latter band, Novices can operate television as well. One major restriction on Novices is transmitter power. In the 80, 40, 15, and 10 meter bands, Novices are limited to 200 watts of transmitter power, 25 watts of power on 1.25 meters, and 5 watts of power on 23 centimeters. Other license classes can generally use transmitter powers of 1500 watts.

Technician: This is the other entry level license. It requires you to pass the Novice written exam and an additional 25 question examination, of which 19 must be answered correctly. *However, the Technician class exam requires no Morse code test!* The Technician class license authorizes you to use all ham frequencies above 50 MHz, with the exception of 50 to 50.1 and 144 to 144.1 MHz which are reserved for Morse code operations only. On other frequencies you can use phone, radioteletype, packet radio, television, and communicate through ham radio satellites. On most frequencies, you're allowed a transmitter power of up to 1500 watts. People who already have a Novice license can add Technician privileges just by passing the 25 question Technician examination. And those who get a Technician license first can add Novice privileges to their bag of tricks just by passing a 5 WPM code test. The unofficial term "Technician-Plus" has been coined to denote those Technician licensees who have passed a 5 WPM code test and have Novice privileges, although the FCC really doesn't issue such a thing as a Technician-Plus license.

General: This license allows all privileges (including phone) in the 160 meter and 10 meter bands, phone privileges in some parts of the 80, 40, 20, and 15 meters, Morse code and radioteletype on most frequencies, and all privileges above 50 MHz. It requires passing a Morse code test at a rate of 13 WPM, a new 25 question written test, and the Novice and Technician tests. You're given full credit for the Novice and Technician exams if you already hold one of those licenses. (When applying for a higher license class, you're given credit for all exams or parts of the exam, such as a Morse code test, you've previously passed even if you don't have a license. For example, suppose you apply for a General class license, pass the 13 WPM code test, but fail the written exams. You'd be given credit for the code test, so next time all you would have to do would be to pass the written exams to earn the General license.)

Advanced: This license gives all privileges allocated to General class licensees plus additional frequencies in the 80, 40, 20, and 15 meter bands, including almost all phone privileges. It requires a 13 WPM Morse code test, the Novice, Technician, and General written exams, plus a new 50 question written test.

Extra: This is the top of the line! It requires passing a Morse code test at a rate of 20 WPM, the Novice, Technician, General, and Advanced written exams, plus a new 40 question written exam. It gives you all Advanced privileges, plus the use of additional frequencies reserved exclusively for Extra class licensees.

For years, examinations for ham licenses were conducted only by FCC personnel. This posed a lot of problems if you lived outside the handful of major metropolitan areas that had FCC field offices. People in the "hinterlands" (which included metropolitan areas with populations in excess of a million!) often had only one or two chances a year to be tested by visiting FCC personnel. (Even then, it wasn't uncommon to have to drive over 50 miles to the test site.) Exams were often held in mid-week during daytime hours, forcing most applicants to miss a day or two of work or school. Finding an opportunity to take a ham license exam at a time that was convenient (or even feasible) was often more challenging than the exam itself!

In response to such problems, and as a cost-saving measure, the FCC in 1985 instituted a program under which already-licensed hams would administer all future licensing exams. These hams are organized into "examining teams" by *volunteer examiner coordinators* (VEC), with all VECs certified by the FCC before they can administer exams. All written examination questions are drawn from a regularly updated pool of questions developed by the FCC and the VECs, and these questions are released to the public. By common agreement of the VECs, the written exams are all multiple-choice, although the correct and incorrect responses are not made public. However, knowing the exact questions you'll be asked on the written exams makes studying for them much easier than in the past!

The method of testing for Morse code varies among VECs. The usual method is to send a five-minute long message, representing a typical ham radio contact, in Morse code. At the end of the message, you're given a ten question test about the message. Some tests are fill-in-the-blank types, while other VECs give multiple-choice tests. If you answer at least seven of the ten questions correctly, you pass. You could also pass the test by copying one consecutive minute of the five-minute message correctly. (Or you could just get a Technician license and say to heck with the Morse code!)

Why is the Technician class the only license that doesn't require a code test? The FCC really has no choice in the matter, since the United States is a signatory to an international communications treaty which requires all persons holding a ham license for operation on the HF bands to have passed a Morse code test. (However, the method to test knowledge of Morse code and the degree of competence required is left up to each nation to determine.) Such a test is not required by treaty for operation on ham bands above 30 MHz, so the Technician class can be issued without a code requirement.

By the way, there's no need to work your way up the licensing ladder. If you're up to it, you could go for the Extra as your first (and only!) license. Some people have done just that. Most hams are like me, however, and start with a lower license class and gradually move up the ladder as they get more knowledge and experience. (I started out as a Novice licensee, and later held the Technician and Advanced classes before finally "topping out" as an Extra.)

VHF/UHF versus HF

The Novice and Technician licensees are both entry points into ham radio, but they follow different paths. The Novice license allows HF (although mainly Morse code), VHF, and UHF operation on a limited range of frequencies, while the Technician license gives access to all VHF and UHF frequencies but no HF operation. And those are two very different worlds.

On the HF bands, it's normally possible to communicate over several hundreds or thousands of miles on a regular basis. As we'll see later, some HF bands are more productive than others for long-distance communications, and they're not always productive at the same time. (For example, 160 meters is best at night during periods of low solar activity, while 10 meters is best during daytime during periods of high solar activity.) But, as a general rule, a ham station equipped for operation on the 160 to 10 meter bands will be able to routinely communicate with other hams thousands of miles away.

Unlike the HF bands, the normal operating range on VHF/UHF frequencies is less than a couple of hundred miles; it's similar to the coverage of your local TV and FM radio stations. Within their normal operating range, however, the VHF and UHF bands provide very reliable coverage.

However, you're not always restricted to local operations on VHF and UHF. In particular, the 6 meter band is capable of DX during periods of high solar activity (these usually occur near the peak of a sunspot cycle). Several hams have worked over 100 countries on 6 meters. In addition, each spring and summer atmospheric conditions permit communications over distances up to about 2500 miles on 6 meters. Other atmospheric conditions (which we'll discuss later) often permit communications over several hundreds of miles on the 2 meter and 70 centimeter band.

VHF and UHF are where most ham radio satellite operations are found. Ham radio satellites are growing in sophistication and importance, and offer hams the same benefits of other communications satellites: reliable, economical global communications. It's possible to conduct satisfactory satellite communications using smaller, less obtrusive antenna systems than for HF communications; as a result, some hams who can operate HF prefer to use VHF and UHF satellite frequencies instead.

Perhaps most importantly, the VHF and UHF ham bands are where packet radio really comes into its own. Packet radio lets you use your personal computer to send messages addressed to specific ham stations and receive and acknowledge messages from other hams even if you're not physically at your station. This is the fastest growing part of ham radio, and it's mainly taking place on VHF and UHF.

There's another difference between the VHF/UHF bands and HF that you should be aware of. As we'll see later in this book, the optimum physical size of an antenna depends upon the operating frequency, and this physical size increases as the frequency drops. It's possible for a highly efficient VHF/UHF antenna to "fit" into a space not much larger than the antenna for your car's radio or the antenna used with a portable TV or FM radio. By contrast, an antenna for 80 meters often requires over 100 feet of clear space. If you live in an apartment, condo, or other area where putting up an outside antenna is restricted or forbidden, VHF/UHF operating may well be the only reasonable option you have.

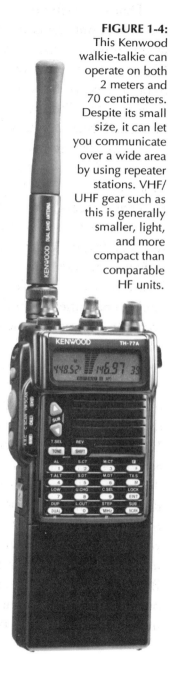

FIGURE 1-4:
This Kenwood walkie-talkie can operate on both 2 meters and 70 centimeters. Despite its small size, it can let you communicate over a wide area by using repeater stations. VHF/UHF gear such as this is generally smaller, light, and more compact than comparable HF units.

Which Should Be My First License?

That's a tough question. For years, the Novice license was the logical starting point in ham radio. However, that was back when all ham licenses (including the Technician) required a Morse code test, and the Novice was the easiest license class to obtain. Now that the Technician class no longer requires a code test, the Technician license is a strong alternative to the Novice. Which one should you try for first?

Many already-licensed hams will probably try to point you toward the Novice class, mainly to get you to learn the Morse code. It's true that a knowledge of Morse code is essential to getting access to the HF bands and a world of fascinating activities, but it's also true that the Novice class license can really be frustrating with its restricted privileges. Novices are only permitted Morse code on 80, 40, and 15 meters, and Morse code can be a slow, frustrating way to communicate if you're a beginner. (Think about it—how long would it take to communicate with someone if you're limited to five words per minute, the code speed requirement for a Novice license? By the way, you're not limited to communicating in Morse at the code speed required for your license. If a Novice can send and receive the code at 20 WPM, then he or she could do so on the air.) Novices are allowed to use phone on 10 meters, but 10 meters is a schizophrenic band. During years of low solar activity, such as 1985 and 1986, the normal communications distance on 10 meters is not much more than that found on the CB channels. Months can go by without hearing any signals beyond a couple of hundred miles of your location. During years of high solar activity, such as 1989 and 1990, it's possible to contact over 100 different countries in a single weekend. But 10 meters does not offer the consistent year-in and year-out international communications ability that a band like 20 meters does. The other two bands where Novices can use phone offer reliable local coverage through repeater stations, but unfortunately these are not heavily used bands.

The big advantage the Novice license has is its access to a part of 10 meters and the possibility of talking to people all around the world using simple equipment. However, Technicians can also talk around the world through ham satellites. You can also exchange written messages with hams across the nation or in foreign countries by using packet radio. The 6 meter band is similar to 10 meters; when 10 meters is good for DX, so usually is 6 meters. Most importantly, the Technician class gives you full access to the 2 meter band, which is by far the most popular ham band. This band is populated by numerous repeater stations to extend the range of portable and mobile ham stations, making it a popular meeting spot for all sorts of hams. Two meters is also where most packet radio activity takes place, and a considerable amount of satellite activity is also found on 2 meters.

Which license is easiest to obtain? For some people, the Novice written exam and the Morse code at 5 WPM is actually easier than the combined Novice/Technician written exam. (This is particularly true for youngsters who haven't yet learned that Morse code is "supposed" to be difficult.) However, most people will probably find the additional 25-question Technician class written exam to be less of a challenge than learning Morse code.

So what license should you shoot for first? Given the greatly expanded scope of operations possible with a Technician class license, you're more likely to get interested in and stay involved in ham radio than if you go the Novice route. The Novice class operating privileges are just too confining for most people; you'll have more fun with a code-free Technician license. You can always get around to learning the code later and add Novice privileges by passing the 5 WPM code test. If you find learning the code easy, shoot for the "Technician-

Plus" license. But my sincere advice is to stay away from the Novice class. For a lousy 25 additional questions, you get access to almost all frequencies and privileges above 30 MHz—and that's a pretty good deal as I see it!

(Am I being hypocritical by suggesting the Technician class as a first license, seeing as I mentioned a few paragraphs ago that my first license was a Novice class? Not at all. . . . back when I first got interested in ham radio, both the Novice and Technician licenses required a 5 WPM code test. The Novice written exam was only 20 questions, while the Technician exam was 50 questions. Back then, the Novice class was clearly the easier route into ham radio. Today, I think the Technician class is easier. Also, back when I got my first ticket, if you didn't upgrade to another license class within two years, your Novice license expired and you were off the air!)

The Morse Code and You

Sooner or later, every ham or potential ham has to face up to the issue of Morse code. Either you learn it or you don't, and even if you do learn it, you might just learn enough to pass the Morse test and then start forgetting it. What should you do?

A friendly suggestion: although I advocate going for a Technician class license as your first one, I also suggest you at least make an attempt to learn Morse code. Why? Some good reasons:

- the Morse code is the key to using the HF bands. While satellites and packet are a lot of fun, it's also a lot of fun to be able to talk directly to somebody on the other side of the world using a HF transceiver! The addition of 10 meter phone privileges is worth the effort of trying to learn code at a 5 WPM level. And you can use an interface device and software with your personal computer to send and receive Morse on 80, 40, and 15 meters.

- it honestly isn't all that difficult. Really! It's nothing more than learning the alphabet, the numbers from 0 to 9, and a few punctuation marks in a foreign language. Could you learn about 45 words in a foreign language? Of course you could. Well, that's all there is to learning the Morse code. I've never been mistaken for a bright person, but I managed to pass the 20 WPM Extra class code test. If I can, you can. Trust me!

- you might wind up enjoying Morse code. Okay, trying to hold a real "back and forth" conversation using Morse can be a pain; it takes forever and there's not a lot of room for spontaneity or a personal touch. But if you're the sort of person who enjoys and treasures the past—perhaps through model railroading or restoring old cars—you might get hooked on Morse. There's something special about being linked to the earliest days of electronic communication through a telegraph key, although today a "telegraph key" is more likely to be an electronic gizmo that makes perfect dots and dashes or even a computer keyboard. In fact, many hams today use special interface devices so they can send and receive Morse code via microcomputer.

So how should you try to learn Morse code? The first place to start is to refer to it as hams do: CW, which is an abbreviation for "continuous wave." This comes from the way Morse code is sent, which is by sending continuous radio waves from a transmitter to form dots and dashes. The second place to start is to forget that "dot" and "dash" stuff, since that's got no relevance to the way the code sounds when sent by radio. Instead, think of *dits* and *dahs*. A dah is three times longer than a dit, and when a dit is sent as part of a series, it becomes *di*. For example, the letter "S" consists of three consecutive dits, and sounds like *dididit*. There's no pause between the individual dits; it's one continuous sound. Table 1-2 shows how the various letters, numbers, and characters of the Morse code sound when sent by radio. Those characters also sound the same way when you're trying to pass the Morse code exam for a license!

TABLE 1-2

Morse Code by Sound Chart

Letter	Equivalent Sound	Letter	Equivalent Sound
A	Didah	U	Dididah
B	Dahdididit	V	Dadididah
C	Dahdidahdit	W	Didahdah
D	Dahdidit	X	Dahdididah
E	Dit	Y	Dahdidahdah
F	Dididahdit	Z	Dahdahdidit
G	Dahdahdit	1	Didahdahdahdah
H	Didididit	2	Dididahdahdah
I	Dididit	3	Didididahdah
J	Didahdahdah	4	Dididididah
K	Dahdidah	5	Dididididit
L	Didahdidit	6	Dahdidididit
M	Dahdah	7	Dahdahdididit
N	Dahdit	8	Dahdahdahdidit
O	Dahdahdah	9	Dahdahdahdahdit
P	Didahdahdit	0	Dahdahdahdahdah
Q	Dahdahdidah	.	Didahdidahdidah
R	Didahdit	?	Dididahdahdidit
S	Dididit	,	Dahdahdididahdah
T	Dah		

Look at Table 1-2 and try spelling out your name in CW. When I do that with my first name, I get *dididit didah didahdit didahdit dahdidahdah*. You could whistle the code or simply pronounce each character to yourself, again using the ratio of a dah being equal in time to three dits. The important thing is to get the sound patterns down in your mind. Don't get into the habit of counting the dits and dahs and then translating them into a character. Memorize the way a character sounds as a whole, so that when you hear *dididahdit* you automatically think of "F."

Try to memorize three or four new characters per day, and practice using them at odd moments throughout the day. When you see a stop sign at an intersection, don't think "stop"— instead, think *dididit dah dahdahdah didahdahdit*. Don't go for a drive in your car; instead, get in your *dahdidahdit didah didahdit* and drive away. All this strike you as silly? I guess it is to a degree, but it's also the way I learned code and wound up with the Extra class license. If you'll give this method a chance, I'm willing to bet that you'll have Table 1-2 memorized in about two weeks. And if you have Table 1-2 memorized, you'll probably be able to pass the 5 WPM code test without difficulty!

Tape cassettes and computer software are available for learning the code and as a source of additional practice material. If you're going to learn the code from tapes or software, beware of those that teach the code as groups of increasing dits and dahs. Such materials might teach the letters E, I, S, H, and the number 5 in the first lesson, and then the letters T, M, O, and the number 0 in the second lesson. The problem with that is you're taught to count the individual dits and dahs instead of recognizing each character by sound. If you count dits and dahs, you'll be stuck around a 5 WPM code speed forever. Once you've learned the characters in Table 1-2, you can sharpen your new skills by copying code from tapes or cassettes. The best tapes and software are those producing random groups of characters; this prevents memorization of the material and the exaggerated idea of your CW ability which can result.

Dididit dit dit dahdahdididahdah dahdidahdah dahdahdah dididdah dah dahdahdah dahdahdah dahdidahdit didah dahdit didahdidit dit didah didahdit dahdit dah didididit dit dahdah dahdahdah didahdit dididit dit dahdidahdit dahdahdah dahdidit dit!

At least make an attempt to learn CW. If you try and discover that you and the code can't get along, well, at least you made the effort. But if you and CW are compatible. . . . why, that could be the start of a beautiful relationship.

Studying for the Written Exams

There are two schools of thought on how to prepare for a written exam on any subject: study to learn the material or study to pass the test. I suggest you study to pass the test. There's plenty of time and opportunity to learn electronics and radio after you get your ham license; there's no need to learn it all in advance.

Don't misinterpret that last paragraph—I'm not saying you don't have to learn any electronics and radio, just that you don't have to become an expert before taking the exam. Since all written exam questions are part of a public pool developed by the FCC and VECs, it's theoretically possible to pass a written exam for a ham license by sheer memorization. But practically it's quicker (and easier) to learn *just enough* electronics and radio theory to pass the written exam. However, you will need to purchase a study guide covering the current exam question pool. You'll need such a study guide to be familiar with the exact questions on the exam and the current FCC rules and regulations you'll need to know.

This book will do its part by conveying just enough knowledge to pass the Novice and Technician class exams. You won't be an electronics expert after reading this book (it's debatable whether you'll even be an electronics "competent" at the end!). But you will be able to pass the exam and have a good foundation for learning more about radio and electronics.

However, you will need to purchase a study guide covering the current exam question pool. You'll need such a study guide to be familiar with the exact questions on the exam and the current FCC rules and regulations you'll need to know. The written exam question pools are updated on a regular basis and are valid for a specified period of time, usually every three years. For example, at the time this book was being written a pool of Technician exam questions valid from November 1, 1989 to October 31, 1992 was in use. After that date, a new set of exam questions will be in use. (However, while the exact questions may change, the topics covered and level of difficulty don't change too much with each revision.) Study guides contain the exact questions of each currently used pool along with five sample multiple-choice answers for each question. While these answers aren't the same as you'll find on the actual exam, they are very similar. Moreover, one of the sample multiple-choice answers will be correct, and it contains the essential idea(s) necessary to answer the actual question. If you memorize each question from the pool and the correct answer given, you should have no trouble with the Novice and Technician written exams.

Exam Study Materials. . . . One Man's Opinion

There are a surprising number of different study guides and packages for the ham exams available. How good are they? Well, the accuracy and technical content of all tend to be excellent, so there's little to choose from in that regard. The choice of which is "best" for prospective hams comes down to factors which can't be quantified easily (if at all). What follows are my admittedly subjective ratings based upon my evaluation of currently available materials. The criteria I've used in this evaluation are numerous and somewhat ineffable, but boil down to one simple "gut-level" judgment on my part: how quickly and easily would a particular study guide or package help someone get ready to pass the exam for the license they want?

#1: Gordon West's Technician Class New No-Code
(Master Publishing, 14 Canyon Creek Village, MS31,
Richardson, TX, 75080)

No one teaches ham radio better than Gordon West, WB6NOA. As a writer myself, I admire and envy his clear, thorough explanations of technical subjects. He stresses teaching enough theory to successfully pass the exam and lays a good foundation for continued learning after you have your ticket. Take this book, add some effort, and a Technician ticket is yours. This book is often available at Radio Shack stores. Gordon also conducts weekend licensing courses throughout the country through his Gordon West's Radio School; write to 2414 College Drive, Costa Mesa, CA, 92626 for more information.

#2: W5YI Marketing Group
(2000 E. Randol Mill Rd., Suite 608-A, Arlington, TX, 76011)
The study materials produced by W5YI are almost as good as those by Gordon West; in fact, there's not a whole lot to choose between them. The W5YI packages also include code tapes and are often available at Radio Shack stores; they are a good choice if you're interested in learning the code as well. W5YI is Fred Maia, who probably deserves more credit than anyone else for starting the movement that finally resulted in the FCC creating a code-free ham license. If you wind up with a Technican class license, and the code was the one thing that kept you from getting a license before, thank Fred.

#3: American Radio Relay League (225 Main St., Newington, CT, 06111).
The League's study guides are strongholds of the "learn it all before you take the test" mentality, reflecting a long-standing "techie" orientation within ARRL leadership. While their materials are terrific if you do want to learn all the ins and outs of electronics and radio before (or after) getting your ham license, they do make the study process more lengthy. However, once you're finished, you know the material thoroughly. The League also offers code learning cassettes which are well done, but—in my humble opinion—these aren't as good as those offered by the W5YI people.

#4 Ameco Publishing Corp. (220 E. Jericho Turnpike, Mineola, NY, 11501).
Ameco's study materials have been around forever, it seems. Unfortunately, they sometimes read and look that way. Production values tend to be low, with many illustrations looking crude and amateurish. Proofreading is sometimes sloppy as well; I've seen a pair of questions and sets of possible answers transposed. (At least, it was easy to figure out from the question and answer content which question went with which set of answers!) On the other hand, there is an adequate discussion of each question, and Ameco's guides are generally the lowest priced of the four mentioned here. Their Morse code course using software for the IBM PC family is very good; their cassette courses are much less so.

If you're really interested in learning radio and electronics, you need to add a good textbook to your ham radio library. My personal recommendation is *Electronic Communication* by Robert L. Shrader, published by McGraw-Hill. The best all-around guide to electronics is *The Art of Electronics* by Paul Horowitz and Winfield Hill, published by Cambridge University Press. And no ham shack is complete without a copy of *The ARRL Handbook for Radio Amateurs* published by the American Radio Relay League. The latter should preferably be a few years old and dog-eared from heavy use!

Scheduling and Taking an Exam

The two main VECs operating nationally are under the auspices of the ARRL and the W5YI group. (There are other VECs, but they are much smaller and offer fewer exam opportunities.) Drop a note to them, along with a self-addressed stamped reply envelope, and ask for the names and addresses of accredited volunteer examining teams in your area. You can then contact your local team(s) directly and determine when they give exams and the procedures you should follow.

Volunteer Exam Coordinators

There are two major volunteer examination coordinators (VEC) that offer examinations for ham radio licenses. To find out when the next exam session in your area will be, drop a request to them along with a self-addressed stamped envelope for their reply.

American Radio Relay League VEC
225 Main St.
Newington, CT 06111
(203) 666–1541

W5YI/VEC
P. O. Box 565101
Dallas, TX 75356-5101
(817) 461-6443

You'll need a copy of FCC form 610 to apply for a ham radio license and to sit for an exam. You can get a copy of form 610 from your nearest FCC office or by writing Federal Communications Commission, Form 610, P. O. Box 1020, Gettysburg, PA, 17326. Don't return the completed form to the FCC; instead, the VEC team that examines you gets it. Some volunteer examining teams, especially those that hold exams at less frequent intervals, will ask that you send your completed form 610 to them in advance of the exam date and schedule an exam time. Other volunteer exam teams offer "walk-in" exams, meaning you just show up with your completed form 610 on the day of the exam. In either case, the volunteer exam team will take your form 610 before the test, certify the exam results (good or bad) on it, and send it to the FCC for processing and issuance of a license.

Volunteer exam teams are allowed to charge a fee, which is set by the FCC, to recover their expenses in preparing and administering exams. Currently, this fee is only a few dollars (less than most movie tickets, as a matter of fact) and the volunteer exam team will notify you of the amount.

When you ask VECs about exams in your area, also ask about the format used for the code tests. Most VECs use fill-in-the-blank type questions, although some use multiple-choice tests. (VECs also are permitted to test applicants on how well they can send Morse code with a telegraph key, although this is almost never done.) If the code test is multiple-choice, you might as well take it even if you don't know a single character of the code. While it's unlikely, you could theoretically pass the code test just by being lucky. (I know one ham who was able to pass the 13 WPM exam that way—he admitted to me he was unable to copy even one letter of the exam message, but guessed right on the multiple-choice questions!) If you've made some attempt to learn the code, try the code test anyway. If you pass, that's great; if not, you're still permitted to try the written exam for the Technician class license. And once you've passed the written exam, you can keep trying the code test at future exam sessions.

You have to pass both the Novice and Technician written exams to earn a Technician license, and both exams are scored separately. In other words, don't forget the Novice class material when studying for the written exam! This might seem a little unfair, since the Novice exam dwells heavily on HF operation and code techniques (at least it did when this book was written), both of which a code-free Technician license cannot utilize. However, that's the way the situation stands and you have no choice other than to live with it.

Have a couple of forms of positive identification with you when you report to your exam session. The best type is some sort of photo identification, such as a driver's license or a passport. You should bring several sharpened #2 pencils or a mechanical pencil to the exam session along with a good eraser. Calculators are permitted, and you should bring one unless doing math in your head is one of your strong points! However, if you bring a programmable calculator, be prepared to clear all memories so as to prevent anyone from showing up to take an exam with all relevant formulas stored in their calculator. There's no need to bring scratch paper, as that will be provided for you.

Maybe it's a little too early to discuss exam-taking strategies, but one should be mentioned here: the best strategy for passing the test is to be prepared. That means you should really work, on a regular schedule, with a current study guide. Try putting in an hour per day of study for a month or so before you take the test. Again, don't get bogged down trying to master each subject completely. Study to pass the test—by rote memorization if need be—and master the details *after* you get your license. Another strategy to remember is that only one completely correct answer will be provided for each question, although some answers might be partially correct. Don't select the first answer that looks correct; read them all and pick the one that is free from obvious error and answers the question most completely. Finally, there's no penalty for an incorrect answer. If you don't know the answer to a question, go ahead and guess. You have a one out of five chance of getting it right even if you don't know a thing about the subject! (And if you're able to eliminate one or more answers as clearly being wrong, so much the better. . . .)

A point that's been made before needs to be reiterated here: the written exam is not as hard as you might think or have heard. Yes, it does require some effort, but that's mainly because the subject matter is likely unfamiliar to most people. The inherent difficulty of the subject matter isn't a problem! If you were able to understand high school science classes, you should have no difficulty passing the written exams for a Technician license. Or, to look at it another way, if you can make it through this book and understand most of it, you *will* pass the Technician exam!

Don't plan to get a ham license "someday." Today is that someday! Start translating words into Morse code, write the FCC for a form 610, get a written exam guide for the Novice and Technician licenses, and find out which VECs give exams in your area. In the meantime, let's keep going and learn the "innards" of ham radio.

Ham Radio Magazines

To get a flavor of what's going on in ham radio, read some of the currently available ham radio magazines. With the exception of *CQ Magazine*, these can be hard to find unless you look at larger newsstands or in ham radio and electronics stores. *QST* is virtually impossible to find on newsstands, although it is available at many public and college libraries. Here's my admittedly subjective review of these magazines; since in the past they have reviewed several of my books, I suppose turnabout is fair play!

QST, c/o American Radio Relay League, 225 Main St., Newington, CT, 06111. The largest circulation of any U.S. ham radio magazine by a wide margin, mainly because a subscription is included with ARRL membership. It tries to be, with varying success, both a newsletter for ARRL members and a general interest ham magazine. Part of the problem with *QST* is that much space in each issue is taken up with news of ARRL elections, summaries of ARRL organizational activity, reports from the various ARRL administrative sections, and related matters that many (including this guy) don't find interesting. There are usually good technical articles in each issue, although they are often at a high level. Since *QST* is the journal of a nonprofit organization, the editorial tone too often tends to be bland, inoffensive, and "institutional." But what the heck? ARRL membership is worthwhile for any ham, and *QST* is well worth your attention.

CQ Magazine, 76 N. Broadway, Hicksville, NY, 11801. My personal favorite. *CQ* stresses DXing, contesting, certificate hunting, and technical articles of a simple to moderate level of difficulty. Many of the DX articles are "first-person" accounts of visits to and operation from rare DX countries such as Cambodia and Tonga. Monthly columns are devoted to such topics as DXing, contesting, packet radio, and operation on the VHF/UHF bands. *CQ* sponsors several major contests each year and major awards, including the tough, prestigious Worked All Zones (WAZ) award. Since this is a commercial magazine, it is more lively and controversial than *QST*.

73 Magazine, WGE Center, Hancock, NH, 03459. The main reason to read this magazine over the years has been the editorials of its founder and editor, Wayne Green, W2NSD. Wayne has long been one of the most controversial people in ham radio, largely because of his fiery editorials. On many issues, Wayne has been the loudest (and sometimes only) voice in opposition to an idea floated by the ARRL or FCC. The rest of the magazine varies in quality according to how much attention Wayne is paying to it compared to his other publishing ventures. When Wayne's concentrating on *73*, the magazine is filled with terrific technical articles, construction projects, and features; when he's busy elsewhere and lets some underlings handle the magazine, the quality usually drops off rapidly and precipitously. What do I think of Wayne Green? No, I don't always agree with his editorials....in fact, I seldom agree with them. But back in the 1960s, when I was first struggling to become a writer, Wayne bought the first magazine article I ever sold. And even when he turned down articles, he sent along letters full of constructive criticism that were among the best writing instruction I ever received. I owe the guy a lot and, for that matter, so does ham radio.

Ham Radio in Canada

Since I'm an American, I've restricted the focus of this book to ham radio as seen from the American perspective. However, there are a lot of similarities between ham radio in the United States and Canada, and much of the information in this book will be useful to Canadians interested in ham radio as well. However, there are some important differences the prospective Canadian ham should be aware of.

In Canada, only one type of license—called the Amateur Radio Operator's Certificate—is issued. However, it comes with four different "levels of qualification," and these levels are *de facto* license classes. The Basic Qualification is required of all Canadian hams, and consists of 100 multiple choice questions on radio theory, operating procedures, and Canadian radio regulations. A passing score is 60%, and obtaining the qualification allows operation on all frequencies above 30 MHz with 250 watts of power; however, home-built transmitting equipment may not be used. The next level of technical exam is the Advanced Qualification, which is 50 multiple choice questions on advanced radio theory. Passing is again 60%, and permits holders to use up to 1000 watts of power, use home-built transmitting equipment, and act as the sponsor of a repeater or club station.

If one passes the Basic Qualification, or even adds the Advanced Qualification, they're restricted to operation above 30 MHz. The Morse Code (5 WPM) Qualification permits operation on 160 and 80 meters on all modes with 250 watts, and requires both receiving and transmitting messages in the code for three consecutive minutes with only five errors or less. The Morse Code (12 WPM) Qualification is the same, but at 12 WPM. The 12 WPM qualification lets you operate on all frequencies and modes below 30 MHz with 250 watts.

These qualifications can be "mixed and matched" as you like. For example, you could obtain the Basic and Advanced Qualifications and be able to operate with 1000 watts of power above 30 MHz. Or you could obtain the Basic and Morse Code (12 WPM) Qualifications and be able to operate on all frequencies above and below 30 MHz, but only with 250 watts of power and using commercially-built transmitting equipment. Another possibility would be to obtain the Basic, Advanced, and Morse Code (5 WPM) Qualifications, which would convey all privileges above 30 MHz and in the 160 and 80 meter bands with 1000 watts of power.

If you're a Canadian, contact the Canadian Radio Relay League, P. O. Box 56, Arva, Ontario, N0M 1C0 for more details about Canadian licensing exams. They also offer several good study guides and references. And I'd like to thank Stephen Canney, VE3FQ, for providing data on the 1990 changes to the Canadian licensing system.

CHAPTER TWO

Ham Radio Customs, Folklore, Habits, and Other Good Stuff

YOUR FIRST EXPOSURE TO HAM RADIO was probably intimidating. All that strange equipment. . . . the weird lingo. . . .those unusual folkways—yeah, it can be overwhelming. Like any other specialized activity, ham radio over the years has built up a set of terms, practices, and customs that "insiders" take for granted. Since they are so important, we'll get to know them in this chapter. We'll also become familiar with the basic equipment in a ham station and how it's all connected together. Some of this material was introduced in the first chapter, but we'll look at it in more detail now.

Call Signs

If you've met many hams, you've probably heard something like this:

"Hi! I'm Harry, AA6FW!"

So what was that all about? Is that "AA6FW" stuff some sort of code, perhaps to identify myself to other secret agents? Or am I trying to tell you I have a really complex blood type?

Neither. That "AA6FW" is something near and dear to every ham—a *call sign* (also referred to as *call letters*, or simply as the *call*). It's issued by the FCC to identify ham stations on the air. Technically, I'm not AA6FW. "AA6FW" is the station license issued by the FCC covering the station at my home, not me. The operator portion of my license is what legally "belongs" to me. In the real world, that's a bogus distinction— I *am* AA6FW! There are a lot of people named Harry in this world, and I've even run across a couple of other people named Harry Helms, but there's only one AA6FW, and that's me. When you get your ham license, your call sign won't just be something to identify your station on the air—it will be *you*.

(When I wrote the above, I thought back to all the hams I've known well over the years. In each case, I had no problem remembering their first name and call sign. However, I could remember the last names of only about 20% of them! That's how closely hams are associated with their call signs.)

Call signs are not assigned at random. They consist of a prefix, a number, and a suffix of one to three letters. Prefixes are divided among nations by international agreement, and a complete prefix list is contained in appendix 1. The United States is assigned the prefixes AA to AL, KA to KZ, NA to NZ, and WA to WZ to identify its ham stations. If you hear a ham station whose call sign prefix is "F," you'll know the station is licensed by France. The same thing goes for "G," which identifies stations licensed by Great Britain.

In the United States, the numeral following the prefix depends upon where you live when the license is issued. For

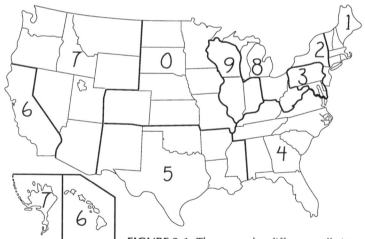

FIGURE 2-1: These are the different call sign districts the United States is divided into. These were set up years ago, and sometimes don't bear too much relation to current population patterns.

example, all persons living in California get "6" as the numeral. If you live in New York or New Jersey, you get "2" following the prefix. The suffix of one to three letters is issued in strict alphabetical sequence. (And no, you can't request a special call sign, like automobile "vanity" license plates, from the FCC.) Figure 2-1 shows these different call sign areas.

Normally, the only valid numbers that can follow a United States prefix are 0 through 9. Sometimes the FCC will permit other numbers to be used by so-called "special events" stations. An example from a few years ago involved the bicentennial of the U.S. Constitution, where the FCC allowed selected stations to replace the usual number in their call sign with 200. But these are rare exceptions, and normally you'll only see 0 to 9 following a prefix.

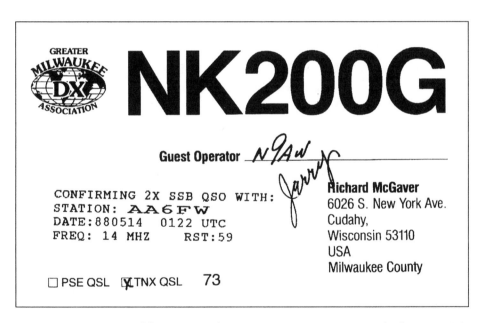

FIGURE 2-2: A QSL card from a special event station commemorating the bicentennial of the U.S. Constitution. In 1988, the FCC allowed certain stations in each state to substitute 200 for the numeral normally used in their call signs for one week as part of the bicentennial activities. As a result, NK9G became NK200G for the period of May 14 through 20, 1988. Some hams (like this one) like to collect QSL cards with "rare" prefixes like NK200. (Yeah, I guess I am pretty weird.)

What happens when you get a license in one call area but move to another, say from New York to California? This happened to me, as I was issued the call sign KR2H when living in New York. When I moved to California, I had two options: I could keep KR2H, or I could ask the FCC to issue a new call sign. At one time, I would have had no choice—the FCC would have issued a new call sign with "6" as the numeral regardless of whether I wanted to keep my old call. A few years ago, the FCC changed its rules to allow hams to keep their current call letters when moving between districts, but with an option of getting a new call if they desire. I'm still a traditionalist who believes that a call sign should reflect where a station is currently located instead of where it was first licensed, so I opted for new call letters. (And if I ever move out of California, I'll again apply for new call letters that reflect where I live!)

Call letters can also indicate the class of license held. So-called "2x1" and "1x2" calls, such as KR2H and W4AA, are held only by Extra class licensees. In addition, the prefix block of AA to AL with a two letter suffix is restricted to Extra class licensees. Advanced class licensees can receive "2x2" calls such as KZ1AA. General and Technician class licensees can get "1x3" call signs, such as W5AAA. And Novices are issued "2x3" call signs such as KZ2ZZZ. Notice how all prefixes are one or two letters? In the United States, all valid prefixes are just one or two letters; there's no such thing as a "3x1" call sign.

There are a few caveats to this neat system, however, mainly based on the limits of the FCC's computer system for processing licenses. (There is no truth to the rumor that the FCC computer was smuggled out of eastern Europe in the early 1960s, however.) Calls are not "recycled" when they become free. Although I haven't held KR2H for several years, it still hasn't been reassigned to another ham nor could I "reclaim" it if I moved back to New York. This means it's possible to not be assigned a call commensurate with your license class if all those allocated to your license class have already been used. For example, if all the

1x3 calls are issued in a district and you qualify for a Technician class license, you would instead be issued a 2x3 call from the Novice group. Nor are you required to change call signs as you upgrade—you have to ask the FCC for a new call if you're eligible for one. If someone likes their Novice call letters, they could hold on to that call even if they upgrade to an Extra class license.

Regardless of the call you wind up with and how you get it, it will soon become your on-the-air persona. And, eventually, you'll start thinking of yourself as "AA6FW" or whatever.

Special Call Signs for Special Places. . . .

There are some variations in the FCC's normal call sign issuance sequence for certain areas of United States territory that are not states. Two states, Alaska and Hawaii, are out of sequence from the rest of the country because they were not yet states when the FCC began issuing call signs using the current system. Whenever you hear one of the following prefixes, you know you're hearing a station—originally licensed, at least—from the following places:

AH0/KH0/NH0/WH0	Mariana Islands
AH1/KH1/NH1/WH1	Baker and Howland Islands
AH2/KH2/NH2/WH2	Guam
AH3/KH3/NH3/WH3	Johnston Island
AH4/KH4/NH4/WH4	Midway Island
AH5/KH5/NH5/WH5	Palmyra and Jarvis Islands
AH6/KH6/NH6/WH6	Hawaii
AH7/KH7/NH7/WH7	Kure Island
AH8/KH8/NH8/WH8	American Samoa
AH9/KH9/NH9/WH9	Wake Island
AL7/KL7/NL7/WL7	Alaska
KC4	U.S. bases in Antarctica
KG4	Guantanamo Bay Naval Base, Cuba
KP1/NP1/WP1	Navassa Island
KP2/NP2/WP2	U.S. Virgin Islands
KP4/NP4/WP4	Puerto Rico
KP5/NP5/WP5	Desecheo Islands
KX6	Marshall Islands

Now admit it. . . . you didn't know there are so many places that are "American territory," did you?

Phonetics

Going back to the opening of the last section, I could have just as easily introduced myself as:

"Hi! I'm Harry, alpha alpha six foxtrot whiskey!"

"Alpha alpha six foxtrot whiskey" means "AA6FW" in the *international phonetic alphabet*. The international phonetic alphabet is "authorized" by the International Telecommunications Union, and is a way of communicating call signs and other information through heavy interference or where one operator may not understand English well. Table 2-1 gives the letters of the alphabet and their equivalents in the international phonetic alphabet.

TABLE 2-1

International Phonetic Alphabet

A	Alpha	N	November
B	Bravo	O	Oscar
C	Charlie	P	Papa
D	Delta	Q	Quebec
E	Echo	R	Romeo
F	Foxtrot	S	Sierra
G	Golf	T	Tango
H	Hotel	U	Uniform
I	India	V	Victor
J	Juliet	W	Whiskey
K	Kilo	X	X-ray
L	Lima	Y	Yankee
M	Mike	Z	Zulu

If you've done any listening on the ham bands, you've probably heard other phonetic alphabets in use. A popular one uses geographic names, as in "America America six Florida

Washington." Another is a variation of the ITU alphabet, using "able" for A, "baker" for B, and so forth. Other phonetics are apparently made up by operators on the spur of the moment. I've been guilty of that in the past; KR2H sometimes came out as "keep running 2 hell"!

But you should memorize the ITU phonetic alphabet in Table 2-1. Why? For one thing, it is the most widely used and understood phonetic alphabet. If someone's native language isn't English, you'll probably find ITU phonetics the easiest way to exchange call signs, names, and locations with them. (Trust me on this; I've worked enough Japanese hams to know.) Cutesy phonetics such as "keep running 2 hell" might work fine with native English speakers or during good conditions, but they can baffle non-native speakers or even other Americans if interference is heavy. Regional accents are another problem with non-standard phonetics—at least with the ITU list, you know what a phonetic letter is supposed to sound like! Finally, the best reason to memorize Table 2-1 is because that is the phonetic alphabet used on the written exams. I can almost guarantee you that at least one question on the Novice written exam will involve "translating" a call sign into the international phonetic alphabet.

Q Signals and Signal Reports

Back in chapter one, we mentioned how some three-letter telegraph abbreviations beginning with "Q" are used in ham radio. These are known, appropriately enough, as Q signals. In the early days of ham radio, these saved all sorts of time and effort by allowing complete thoughts to be communicated using only three letters. (Not a bad idea for everyday conversation today, come to think about it!) They're still used for that same purpose, but have now become part of the everyday lingo of ham radio. Table 2-2 has a list of the most commonly used Q signals; these are also the ones you're likely to face on the written test.

TABLE 2-2

Commonly Used Q Signals

QRS   Send more slowly

QRT   Stop sending or transmitting

QRL   The frequency is busy (often asked as a question before transmitting on a given frequency to determine if it is in use)

QRP   Reduce transmitting power (used generically to refer to any low-power transmitter)

QRX   Please wait

QRZ?   Who is calling me? (almost always used as a question)

QSL   I acknowledge receipt of your message

QTH   My location is. . . .

You can turn any Q signal into a question just by following it with a question mark or by asking it as a question on phone. Many of these signals are informally used as nouns or verbs, as in "I QSO'd him last night on 75" or "I finally got that QSL (card) from Rwanda."

To tell other stations how well their signals are being received at your QTH, the *RST system* is used. "RST" stands for readability, strength, and tone. "Readability" means how easily you can understand the signal, and is rated on a scale from 1 to 5, with 5 being a perfectly readable signal and 1 being unreadable; a 5 is a clear, strong signal like that from a local AM or FM broadcast station, while a 1 means you can tell a signal's there but you can't understand what you're hearing. "Strength" means how strong the station's signal is, and is rated from 9 (as strong as a local AM or FM broadcast station) to 1 (barely audible). Doesn't the "readability" cover signal strength as well? Not really; a strong signal can suffer from heavy interference, for example, or a station can be really weak but you can understand every word. The final rating, "tone," is used only with CW signals and refers to how close the CW comes to sounding like a pure, steady tone, which gets a 9 rating. As the rating drops, say to a 5, the CW signals sounds more like a buzzer than a tone. At the bottom of the scale, the CW sounds like bursts of noise. (We'll discuss why a CW transmitter might put out something less than a "T9" signal later in this book.) If a signal report is being given to a phone station, only the "RS" portion of the report is sent. Table 2-3 gives the meaning of the RST system.

TABLE 2-3

The RST Signal Reporting System

Readability	1	A signal is "there" but is unreadable
	2	Words or characters can be detected, but much of the transmission is missed
	3	Readable with difficulty; most of the transmission can be understood
	4	Readable with only a little difficulty
	5	Readable without difficulty
Strength	1	Signal just barely audible
	2	Very weak signal
	3	Weak signal
	4	Fair signal
	5	Medium signal
	6	Good signal
	7	Moderately strong signal
	8	Strong signal
	9	Powerful, "local quality" signal
Tone	1	Very rough, noisy sound—painful to hear!
	2	Rough, noisy sound
	3	"Buzzing" sound
	4	Smoother "buzzing" sound
	5	Rough, varying tone without "buzz"
	6	Slightly smoother varying tone
	7	Tone varies in a steady pattern
	8	Slightly varying tone
	9	Pure, steady tone

The RST system isn't exact. A signal I might rate as "57" might instead be rated "59" or "55" by other hams. This is why the system is often treated as something of a joke during contests or DXing; everyone gives each other "59" (spoken as "five by nine") reports even when they have to ask each other to repeat their call signs four or five times! Yet the RST system endures and probably will forever. Try to give the most accurate reports you can, but don't take all the "59" reports you get from other operators too seriously!

Making Contact

How do hams make contact with each other on the air? Let's suppose I want to talk to any ham that hears me. The first thing I do is search for a frequency that's not being used by other hams. When I find one that seems empty, I make sure by asking "is this frequency in use?" (On CW, I'd send "QRL?", which means the same thing.) If I hear no reply, I then say something like the following:

"Hello, CQ. Calling CQ, CQ, CQ. . . . this is AA6FW, alpha alpha 6 foxtrot whiskey calling CQ. Hello CQ, CQ, CQ, CQ, calling CQ. This is alpha alpha 6 foxtrot whiskey, AA6FW, calling CQ, CQ, CQ, CQ. This is AA6FW calling CQ and listening."

"CQ" is a general call to any station listening and indicates I'm willing to answer anyone who hears me. If no one answers me, I repeat the same sort of CQ call until someone does.

Notice that my CQ call isn't that long, taking less than a minute. Some hams, particularly newer ones, make *looonnnnggggggg* CQ calls on the theory that more people might hear them. Actually, most other hams won't have the patience to wait for a long-winded CQ caller to shut up long enough for an answer to be sent. The best way to call CQ is to try shorter calls more frequently. If you don't hear an answer within 30 seconds or so after calling CQ, then try again.

On CW and radioteletype, an equivalent CQ would be like this:

CQ CQ CQ DE AA6FW AA6FW AA6FW
CQ CQ CQ DE AA6FW AA6FW AA6FW
CQ CQ CQ DE AA6FW AA6FW AA6FW K

with "K" being the CW way of saying "I'm now listening" (among other things).

To reply to my CQ, you could say something like "AA6FW, AA6FW, this is KR2H, kilo romeo 2 hotel, how copy?" on phone or "AA6FW AA6FW DE KR2H KR2H AR" on CW or RTTY, with "AR" indicating the end of the transmission. I would then acknowledge your call by saying something like "KR2H, thanks for the call, your signals are a solid five by nine here in. . . . " and we'd start having a normal conversation (or at least something vaguely approximating one).

You might have the notion that operating procedures during a contact are really complex, but they're not. It's usually just like a conversation you'd have on the telephone, except that you and the other ham(s) have to take turns speaking. In the old days, you had to give the call sign of the station you were in contact with as well as your own with *each* transmission you made. Today, however, you just have to identify your station at least once every ten minutes ("This is AA6FW for ID") and at the end of the contact. There's no need to identify the station you're in contact with ("KR2H, this is AA6FW") although you'll sometimes hear it done; a simple "this is AA6FW" is all that you need.

Sometimes you'll hear two or more stations using the same frequency, as when a group of friends get together on the air. If it's an informal gathering, this is known as a *roundtable*. At other times, you can hear stations sharing a frequency in a much more formal manner, and this is known as a *net*. Nets have a *net control station* that directs their operation. Stations formally check in a net and are recognized by the net control station; stations only transmit when directed to do so by the net control station and must request permission of the net control station to leave the net. The reason for such tight discipline is to prevent chaos in an actual emergency; many nets are established to provide emergency communications, and everyday use of rigid operating procedures makes them second nature in a real emergency when events are moving fast and furious.

These procedures for establishing contact and conducting communications are good general rules and are followed in phone and CW communications. However, as we'll see in the very next section, specialized methods of communications have their own specialized procedures.

The American Radio Relay League (the ARRL, a.k.a. "the League") and You

The ARRL is the only true national association representing ham radio operators, and that's been true since 1914. Today the League provides a wide range of services of benefit to ham radio operators. These include their monthly magazine *QST*, a large number of very useful books about ham radio, sponsorship of a number of contests and operating awards (including the popular "DX Century Club" award for working 100 or more different countries), news bulletins about ham radio topics transmitted from their headquarters station W1AW, sponsorship of traffic handling networks, and having a significant input into FCC decisions involving ham radio. About one-third of all licensed hams are ARRL members.

The relationship between the majority of hams and the ARRL has not been entirely smooth over the last three decades, however. While the ARRL is nominally democratic (governed by a board of directors elected by members on a regional basis), the ARRL has found itself on the wrong end of the sentiment of the majority of hams (including non-ARRL members) on a couple of key issues. One was the so-called "incentive licensing" controversy in the mid-1960s. Thirty years ago, the General class license conferred all amateur frequencies and privileges, the same as the Extra class license does today. (Back then, the Advanced class license was not being issued and the Extra class license conveyed only glory.) A small group of ARRL leaders felt that it was important to do something about what they felt was deteriorating standards of operation and knowledge among hams, and their solution was to require hams to obtain an Extra class license to secure those privileges formerly conveyed by the General. Unfortunately, the ARRL leadership at the time reached this conclusion in secret instead of through consultation with the membership, and as a result submitted a draconian proposal to the FCC that would have banned HF phone operation altogether by General class hams on 80, 40, and 20 meters unless they upgraded to an Advanced or Extra class license. Unfortunately again for most hams, the FCC took the ARRL proposal seriously and created substantially the license class and operating privilege structure we have today. The growth in the number of new ham licensees stopped dead in its tracks when the FCC adopted incentive licensing, resulting in problems that ham radio (the low rate of growth) and the ARRL (a sullen membership and often hostile non-members) were still wrestling with years later.

More recently, the ARRL leadership was hostile to the notion of any sort of code-free license. In both 1975 and 1981, the FCC proposed creating a code-free license class, and both times the ARRL filed strong comments in opposition. However, by 1989 the number of new persons obtaining ham licenses had dropped to critically low levels, and the long term viability of ham radio was threatened. A group of ham radio equipment manufacturers and others (including Tandy Corporation, parent company of Radio Shack) petitioned the FCC for some form of code-free license. The FCC reacted favorably and asked for public comments. The ARRL's comments in "support" of a code-free license were lukewarm at best; the ARRL wanted to keep code-free licensees off the popular 2 meter band (virtually assuring that such hams would forever be isolated from the mainstream of ham radio), enforce lower power levels for them, and still include some questions about Morse code on the written exam. Unlike the incentive licensing debacle, however, the FCC largely ignored the ARRL comments and instead followed the comments of others which urged simply dropping the existing 5 WPM code requirement for a Technician license.

Given this history, should you join ARRL? Damn right you should!

The ARRL has traditionally been governed by die-hard traditionalists who were reflexively opposed to change of any sort. However, the ARRL headquarters and QST magazine staffs in recent years have been far more closely aligned with the thinking of the majority of hams. Now much of the leadership of the ARRL has come around to accept the need for a code-free license and a new approach to some of the problems facing ham radio. The result is that today's ARRL is undertaking exciting new initiatives and programs that would have been unthinkable less than a decade ago. And the ARRL still provides services that are essential to the very survival of ham radio.

You might hear some hams grumble about the ARRL over incentive licensing and their opposition to a code-free license. Yeah, those were bonehead moves, but that was then and this is now. Without the ARRL, ham radio would have been carved up and sold off by special interests a long time ago. A healthy League is essential for healthy ham radio. A membership in today's ARRL is probably the best investment you can make in getting the most enjoyment out of your new hobby. Drop them a line at ARRL, 225 Main St., Newington, CT, 06111 and ask for membership information and an application form. You don't have to have a ham license to be a member, although full voting privileges are extended only to licensed hams.

Repeater Stations

Many hams do a lot of operating with the assistance of another ham station which automatically receives and retransmits their signals for improved coverage. A repeater listens for signals on one frequency, known as the *input*, and simultaneously relays them on a second frequency, known as the *output*. Stations using a repeater transmit on the repeater's input and listen on the repeater's output; this process of transmitting on one frequency while listening on another is known as *duplex* operation. (Transmitting and receiving on the same frequency is known as *simplex* operation.) It's accepted practice to refrain from transmitting on repeater output frequencies. Not only does this cause unnecessary QRM, but the repeater will "squash" any simplex QSO on the output frequency when it comes on! Frequencies like 146.52 MHz have been set aside for simplex use in band ranges where repeaters are found. Most repeaters use *frequency modulation* (FM), the same transmission method used on the 88 to 108 MHz FM broadcast band.

By agreement among hams, certain pairs of frequencies have been set aside for input and output use, much like TV channels. Current FCC rules allow repeater operations on 10 meters and higher bands, with the heaviest activity on 2 meters. If you have a "scanner" radio that can tune in your local police and fire departments, you can probably hear several repeaters in your area on it and get a sampling of what this aspect of ham radio is all about.

Listening to FM Repeaters

The 5 kHz deviation used for ham radio FM repeaters is the same used by other FM two-way radio services in the VHF/UHF range. If you have a programmable "scanner" receiver that tunes 144 to 148 MHz, you can listen in on ham repeaters on the popular 2 meter band. Try programming the following FM repeater output frequencies into your scanner and see which ones are active in your area:

145.11	145.35	146.73	147.09
145.13	145.37	146.76	147.12
145.15	145.39	146.79	147.15
145.17	145.41	146.82	147.18
145.19	145.43	146.85	147.21
145.21	145.45	146.88	147.24
145.23	145.47	146.91	147.27
145.25	145.49	146.94	147.30
145.27	146.61	146.97	147.33
145.29	146.64	147.00	147.36
145.31	146.67	147.03	147.39
145.33	146.70	147.06	

In addition to these repeater output frequencies, 146.49, 146.52, and 146.55 MHz are often used for direct communications without a repeater (simplex). Also, if you're curious as to what packet radio sounds like, 145.01, 145.03, 145.05, 145.07, and 145.09 MHz are common packet radio frequencies. You'll be able to recognize packet by its distinctive "brapp–brapp!!" sounds.

Operating procedures on a repeater are different from those we looked at before, mainly because many stations continuously listen to the output frequency and clearly hear anything transmitted on the input frequency. Lengthy CQ calls are out of place. Instead, a simple "this is AA6FW listening" is enough to initiate a contact on a quiet repeater "channel." If you want to establish contact with a specific station that you think might be listening, just say "KR2H, this is AA6FW. Are you around?"

Repeaters are often busy, and you might want to (or need to) join a conversation in progress. To help accommodate such situations, it's good practice for each station using a repeater to pause a couple of seconds between transmissions. This allows stations desiring to use the repeater to do so by slipping in their call signs between transmissions. The next station to transmit then acknowledges the new station, and allows it to make a call. *Don't* try to gain access to a repeater by saying "break" or similar CB lingo between transmissions; doing so is the mark of a real yahoo.

Repeaters are terrific at increasing the range of hand-held and mobile stations and allowing stations scattered across a wide geographic area to communicate effectively. They're also a real boon to stations that can't erect efficient antennas (such as those hams living in apartments and condos). A neat feature of many repeaters is autopatch. An autopatch is a connection between the repeater and your local telephone network which permits telephone calls to be made using the repeater. Many mobile and hand-held two meter transceivers include a telephone tone dialing pad to permit use of a repeater autopatch. These calls are simplex, not duplex; you and the party called have to take turns talking and listening. Moreover, since the call goes through the repeater, anyone listening on the repeater output frequency will be able to hear both sides of the conversation (something that many non-hams who are called via autopatch often are unable to grasp!). While it's not as convenient as a cellular phone—and you can't use it for business purposes—autopatch facilities are a valuable addition to any repeater, especially for emergency use.

Repeaters are usually sponsored by a ham radio club; in a few cases, they are also operated by individuals. Most repeaters are *open*, meaning any properly licensed ham can use them just by transmitting on the repeater's input frequency. A few repeaters are *closed*, meaning their use is restricted to members of a particular club or group supporting the repeater. Access to a closed repeater usually involves having to transmit a certain tone (often outside the range of normal hearing, or *subaudible*) at the start of each transmission to "unlock" the repeater's receiver. Some open repeaters also require tones to access them, but this is to ensure that a particular station really wants to access the repeater. For open repeaters, any required access tones are public knowledge.

Even if a repeater is open, it's a good idea to join the club or other group that financially supports it if you're a regular user. The costs of installing and maintaining a repeater are significant, and it's only fair to do what you can to support those repeaters you use.

Repeater frequencies and operations are facilitated by *repeater frequency coordinators*. These volunteers, similar to VECs, assign frequencies for repeaters on a regional basis to prevent interference between repeaters using the same input and output frequencies.

Digital Modes

The microcomputer revolution has left its mark on ham radio too. There are a lot of hams today who spend most of their time communicating with a personal computer instead of a microphone or key. Any method (called a *mode* by hams) of transmission that involves using a microcomputer is usually referred to as a "digital mode."

In case you're not familiar with what "digital" means, it refers to a signal that can assume just two possible conditions or states. For example, a signal can be either present or not present (or "on" or "off"). In a sense, a flashlight is a "digital" device since it can assume just two states: on and off. Morse code is also a crude digital mode, since it involves just the "dit" and "dah" states. Modern digital communications using microcomputers involve switching back and forth between two different radio frequencies or two different audio tones (much like how a computer modem works, if you're familiar with them).

A popular digital mode is *radioteletype* (RTTY). In RTTY, the transmitted frequency shifts back and forth between two separate frequencies. The higher frequency is called the *mark* frequency while the lower is known as the *space* frequency, and on the ham bands these are separated by 170 Hz. This shifting is used to generate characters in the *Baudot* code representing letters, numbers, and punctuation. The Baudot code uses various combinations of marks and spaces to form different letters, numerals, and punctuation; all Baudot characters are five "bits" (that is, individual mark or space signals) in length. At the receiving station, the Baudot characters are decoded and used to display printed messages on a video screen or printer. RTTY is still very popular on the HF bands.

If you're a personal computer user, you already know about the American Standard Code for Information Interchange (ASCII). ASCII is a seven-bit code used to store data on computer disks or send it via modem to other computers. ASCII can be sent via ham radio in the same way as RTTY. The two additional bits used in each ASCII character allows ASCII to have upper and lower case letters and a wider range of symbols (such as <, %, etc.). ASCII lets you transmit computer software and data files via ham radio.

A problem with RTTY and ASCII is they have no way to automatically check whether or not a message is being received correctly. Microcomputers permit the use of modes with error detection and correction capabilities. A popular one among hams is *amateur teleprinting over radio* (AMTOR). AMTOR also uses a seven-bit code like ASCII, but gives two methods of error detection and correction. For transmissions directed to no station in particular (such as a CQ call), each character in an AMTOR message is sent twice. On the receiving end, the message is checked for accuracy and the extra characters are discarded if not needed. Once contact has been established between two stations using AMTOR, they can be "linked" together during a QSO and rapidly switch between receiving and transmitting to acknowledge whether or not a message was received correctly. If it was, the receiving station sends an acknowledgment to the transmitting station; if not, the receiving station will request a repeat of the message from the transmitting station.

The latest advance in digital communications is *packet* radio, which is a lot like an on-the-air version of landline computer networks such as Usenet, CompuServe, and GEnie. This has even more advanced error detection and checking features than AMTOR, and allows features such as automatic routing of messages to their destination and bulletin-board services. Suppose you're located in Boston and want to send a message to a ham friend in Los Angeles. If you're both packet radio users, there's no need for direct contact. You can "launch" the message from Boston, where it will enter the packet network and be relayed automatically by other packet stations across the country until it reaches Los Angeles. Your friend can then check a packet radio "bulletin board" in Los Angeles and retrieve your message. These packet radio bulletin boards are much like their landline cousins; you can read messages on them, post messages on them, and download files (including computer software).

Packet radio also allows fully unattended operation, so your friend could arrive home and find your message waiting for him or her in the personal computer used in their packet station. Fully unattended operation of a packet station lets your packet station receive and transmit messages when you're not around, or even transmit an "answering machine" message to other hams who try to contact you when you're not available.

Packet networks make extensive use of *digipeaters*. Digipeater is short for "digital repeater." Like voice repeaters, digipeaters receive and relay transmissions to other stations. Digipeaters are much simpler than voice repeaters, since virtually any packet radio station can be operated as a digipeater by a few simple software commands. One advantage of digipeaters over voice repeaters is that packet messages have to be specifically addressed to a digipeater to be relayed, otherwise the digipeater will "ignore" the message. In contrast, a voice repeater relays everything (including noise and interference) it hears on its input frequency.

Packet radio is a very exciting development in ham radio, but is still very much in its infancy and there's plenty of room for experimentation and new developments. We'll take a closer look at packet radio later in Chapter 4.

FIGURE 2-3: This is what makes packet radio possible—a terminal node controller (TNC) that ties your computer and ham radio transceiver together. We'll take a closer look at packet radio and TNCs in Chapter 4.

Video Communications

If talking or typing at a computer keyboard becomes boring, how about showing your smiling face to fellow hams? Novice class licensees are allowed limited television privileges, while Technician class licensees are allowed full television privileges.

There are actually two types of television hams can use. One is *slow-scan television* (SSTV), which is a system by which still pictures can be transmitted on HF frequencies. In SSTV, a still television picture is converted into audio tones so it can be transmitted in the same amount of frequency space used for normal phone communications. The television we're most familiar with (namely, the type with a picture that moves) is called *fast-scan television*. However, fast-scan television requires much more frequency space, typically 4 MHz, and so is permitted only on ham frequencies above 420 MHz. The reason why fast-scan television needs so much more frequency space is that a moving television picture contains more "information" in it than a still picture, and a wider frequency range (or *bandwidth*) is needed to contain the additional information.

SSTV was a big thing when it was first introduced about twenty years ago, but has not grown much in popularity since then. Fast-scan television, by contrast, is now coming into its own, particularly in urban areas where some television repeaters have been set up. The availability of television cameras, video recorders, and transceivers for ham TV frequencies have made it much easier to get on the air with fast-scan television, and this area of ham radio is enjoying some exciting growth.

Satellite Communications

Satellite dish antennas are a familiar sight in the backyards of homes in rural areas. Hams communicate with each other through satellites, and they don't need antenna systems as big or elaborate as satellite television viewers need!

One way to think of ham satellites is as orbiting repeater stations. The satellites used by ham radio operators aren't the same ones used to relay television signals or telephone calls. Instead, hams have built their own series of satellites and orbited them, with the assistance of cooperative governments, as "piggyback" payloads on regularly scheduled launches. The best known ham radio satellites belong to the OSCAR (orbiting satellite carrying amateur radio) series, and additional satellites have been built and orbited by Japanese (the "Fuji" series) and Russian ("Radio Sputnik") amateurs. The very first OSCAR satellite was launched in late 1961. This satellite did not relay ham transmissions, but instead sent the letters "hi" in Morse code. Hams all over the world received OSCAR I's signals, proving that it was possible to receive low-powered signals from orbiting satellites with simple equipment.

In 1965, OSCAR III was launched to relay signals from ground-based ham stations. During its 18 days of operation, over 1000 hams in 22 countries made contacts through it (including contacts between hams in the United States and Europe). The basic concept behind OSCAR III has been followed by all subsequent ham satellites. OSCAR III used a *transponder*, which is a device that will listen for input signals within a certain frequency range, such as 50 kHz, and retransmit them within another 50 kHz range. A transponder is different from a repeater in that a transponder receives and retransmits everything within the frequency range of the transponder while a repeater only receives and retransmits a specific frequency; a transponder can relay more than one station at a time. Later in 1965, OSCAR IV was launched. This also carried a transponder, and it allowed a ham in the United States and the Soviet Union to make the first satellite contact of any type between the two countries.

Today, there are two main types of ham communications satellites, and they carry multiple transponders for different bands. One is known as the *low orbit* or *Phase II* satellite. These satellites travel in relatively low (a couple of hundred miles or so) orbits and can be accessed from the ground for only a limited time, less than 20 minutes per orbit. The maximum possible communications range through a low orbit satellite is about 3000 miles. The other major type is the *Phase III* satellite. Such satellites have highly elliptical orbits, reaching a maximum altitude of 22,500 miles, and take about 11 hours for each orbit. Because of the higher altitude, a Phase III satellite "sees" more of the Earth and allows communications over a much greater range. For example, ham stations in Hawaii, Alaska, Miami, Newfoundland, and Brazil could all communicate with each other simultaneously through a Phase III satellite located over the Pacific coast of Central America.

Ham satellite transponders listen for input signals on one band and output those signals on a separate band. For example, signals from ground stations to the satellite (the *uplink* signals) might be transmitted near 435 MHz; the transponder aboard the satellite might then relay signals back to the ground (the *downlink* signals) on a frequency near 146 MHz. Other popular uplink/downlink transponder ranges are 146/29.5 and 1269/436 MHz.

Most ham radio satellites are the low orbit type. These have increased in sophistication so that it's now possible to uplink packet messages to them and have those messages downlinked as the satellite orbits to other stations beyond direct communications range of the originating station.

Strictly speaking, the chance to talk to someone aboard the Space Shuttle or Mir space station isn't "satellite communications." But a number of hams around the world have had the chance to do just that! Several of the American and European astronauts, as well as Russian cosmonauts, have ham tickets and have operated on VHF (usually 2 meters) from Earth orbit. Very modest stations have accomplished this feat (in fact, it's been done with walkie-talkie VHF units!). Future operating plans of "hams in space" are widely publicized in ham magazines.

Third-Party Traffic

"Third-party traffic" refers to any message passed from one ham station to another for someone other than the operators of the two stations. Suppose you're helping provide communications for a marathon race, and use a hand-held transceiver to report results from a checkpoint to the race control center. Those results are third-party traffic. If you let a non-licensed friend say "hello" over your station, you're passing third-party traffic—the "hello" from your non-licensed friend.

In order to prevent ham radio from turning into another band for taxis or businesses, there are several major restrictions on third-party traffic. As mentioned earlier, you can't receive money for any third-party traffic you handle or for operating your ham radio station. And you can't operate your ham station so that someone receives a direct commercial benefit from your doing so. This means that you can't use your ham radio to help you run your business or even to order a pizza using a repeater autopatch.

(There is one very obscure exception to these restrictions. A person may be paid to serve as the operator of a club ham radio station when that station is primarily used to transmit code practice material or information bulletins for the amateur radio service. This loophole allows W1AW, the "club station" of the American Radio Relay League, to pay its operators to send—sonofagun!—code practice material and information bulletins. For several years, the ARRL had paid the operators of W1AW until someone noticed that the practice was illegal. The League quickly petitioned the FCC to allow some form of payment to its operators, and the FCC went along. The rule is so narrowly written that it's unlikely that anyone other than the ARRL could ever qualify.)

If you let someone speak over your ham radio station (or operate the computer keyboard in packet radio, etc.), you must be physically present and operate the transmitting equipment yourself. In other words, you can't let friends without licenses use your ham radio station all by themselves or "loan" the privileges of your ham license. You must be present to supervise all operations, make all adjustments to the transmitting equipment, and ensure that all FCC regulations are fully obeyed.

Within these restrictions, you can send all the third-party traffic you want between your station and other ham stations in the United States. Such isn't the case when you exchange third-party traffic with ham stations in foreign countries. In many nations, telecommunications is a government monopoly and those governments don't take the loss of potential revenue lightly. As a rule, you can't exchange third-party traffic with a ham station in a foreign country unless there is a *third-party agreement* between that country and the United States. Most of these countries are in the Americas (such as Canada, Mexico, Brazil, Guatemala, Jamaica, Costa Rica, and even Cuba) along with an interesting scattering of countries like Australia, Israel, and Swaziland. The latest list of these countries is published periodically in magazines such as QST. If you're in doubt whether you're permitted to pass third-party traffic to a foreign country, the safest course of action is to decline unless it's an emergency.

The ARRL, appropriately enough for an organization with its roots as a "relay league," has an extensive organization to facilitate the origination, passing, and delivery of third-party traffic within the United States. Several on-the-air nets under the auspices of the ARRL meet daily to handle traffic, using phone, CW, and digital modes. A few hams even make "traffic handling" their major operating activity.

Emergency Operations

We've noted earlier how valuable ham radio is during emergency situations. The FCC recognizes this, and has provided special provisions in the regulations for the amateur radio service to facilitate emergency communications.

One invariable rule is that emergency communications have priority over all other communications. If you hear an emergency call on a frequency you're using, you must immediately cease normal communications and stand by to copy the emergency message and render all assistance you can to the station making the emergency call. You also must avoid causing interference to any emergency message, such as operating on a frequency adjacent to one with emergency communications and causing interference to emergency traffic.

If there is a situation in which normal communications systems have been disrupted (such as an earthquake, hurricane, or other natural disaster) and human life is in immediate danger, hams can use any means of radio communication at their disposal even if those means are not authorized by FCC rules or their class of ham radio license. Suppose you hold a Technician license and happen to hear a ship in distress calling for help on a frequency in the HF bands where Technicians are not authorized to use voice. If no other station answers the ship's call, you can reply to the distress call even though Technicians are not normally allowed to operate on that frequency. Ham stations can also communicate with non-ham stations, such as civil defense and government stations, during an emergency. These exceptions are allowed only during *bona fide* emergencies, such as 1989's San Francisco earthquake and Hurricane Hugo. Airplane crashes, tornadoes, and floods are emergencies; someone running out of gas on the highway isn't. A sinking ship is an emergency, while a ship with engine trouble but in no danger of sinking wouldn't qualify as an emergency. However, the latter situation could turn into an emergency if the ship were unable to summon assistance using its marine radio or if bad weather developed.

Formal emergency communications are under the auspices of the Radio Amateur Civil Emergency Service (RACES), which is a part of the Federal Emergency Management Agency (FEMA), the U.S. government agency responsible for coordinating the federal response to major disasters. Hams desiring to participate in RACES must be registered with a civil defense organization which operates RACES facilities. RACES stations can only be used for civil preparedness communications and drills, and drills are limited to only one hour per week.

If a disaster disrupts normal communications facilities in an area, the FCC can declare a *temporary state of communication emergency*. This declaration outlines any special rules or conditions to be observed during the communications emergency, and is intended to facilitate relief actions and help meet essential needs in the disaster area.

Not all communications in emergency situations are of the same importance. The most important is *emergency* traffic, which has life-or-death urgency. Evacuation of injured persons, requests for medical help or critical supplies, and reports of the movement of a tornado are all examples of emergency traffic. Less important is *priority* traffic, which is still urgent but not life-or-death critical. Communications relating to the safety of property (instead of people) often fall into this category. The least important traffic is *welfare*, which relates to the safety and status of people in the disaster and damage to property. Welfare traffic asks whether someone or something in the disaster area is okay, and is how people in the disaster area let others know they or their property is alright. Communications in an emergency situation should be handled in this order of priority. While certainly many people want to know how their relatives are doing in a disaster area, those communications are clearly not as important as those which help get essential services and supplies to those still in the disaster area.

With the spread of repeater stations and improved local communications capability, an increasing amount of emergency communications are of the *tactical* variety. Tactical traffic is an informal, immediate response to a situation within a small geographic area, and is composed of messages such as "we have an injured person over here." In tactical communications, so-called *tactical call signs*, such as "Red Cross" or "emergency headquarters," may be used to indicate which stations are associated with various facilities or services. This is particularly useful if hams are working with non-hams in an emergency situation. (In fact, tactical call signs are becoming popular for non-emergency situations, such as marathons and parades, for this reason.) Use of tactical call signs *does not* excuse stations from having to identify themselves using their FCC call signs each ten minutes and at the end of their communications; tactical call signs are a supplement to FCC call signs, not a substitute for them.

Good emergency communications depend upon hams being prepared for emergencies. Having some portable radio equipment that can operate from batteries or your car's electrical system is a good start. The best type of antenna for emergency use is a simple one that can be transported and erected easily. The most common type is known as a *dipole*, and will be discussed later in this book.

You Can't Say That on the Radio!

In addition to the prohibited activities (broadcasting, etc.) mentioned before, here are some other things the FCC says you can't say or do (whether by voice, Morse code, or whatever) over your ham radio station:

- False or deceptive communications, such as fake emergency calls

- Unidentified communications, in which you don't use your FCC-assigned call sign or don't identify as required by FCC rules
- Deliberately or maliciously cause interference to another station
- Obscene, indecent, or profane words or meaning

The last two items need some elaboration. The FCC doesn't prohibit causing any interference, only *deliberate* or *malicious* interference produced in a willful manner. Many HF ham bands are crowded, and some interference—which can be heavy at times—is often unavoidable. Except for emergency traffic, no ham station has a "right" to any frequency. If you're in a contact, and changes in conditions allow you to hear an interfering station on the frequency, you have no right to order the interfering station off the frequency. By the same token, if you have a schedule with a friend set for a certain time and frequency, you have no right to tell stations already using that frequency at the appointed time to move. However, a little common sense and courtesy can resolve most interference problems. The other operators want to be rid of interference as much as you do, and a clear frequency can be usually found with a little effort. If you're suffering interference, take the first step and offer to move to a clearer frequency. If someone's using a frequency you have a schedule with a friend on, ask if you can call your friend on that frequency when you find a new, clear frequency. Most hams will try their best to help if asked politely; most hams will also dig in their heels if they think they're right and someone's trying to order them around!

What are "obscene, indecent, or profane words or meaning"? Your guess is as good as mine. An occasional "hell" or "damn" doesn't qualify, but beyond that the issue is murky (let's face it, even the Supreme Court hasn't been able to precisely define what is or isn't obscene). But unless your communicative powers are badly restricted and/or you're not too bright, you can probably express yourself quite well without resorting to words that might be considered obscene, indecent, or profane. If that's the case, try leaving them unspoken. The key to making ham radio fun for everybody is to avoid making yourself a pest or problem, so why offend someone needlessly over something trivial?

By the way, you might run into some hams who say you should avoid discussing controversial topics such as religion and politics on the air. Such people sound like the sort of cranks who are unable to disagree with someone without becoming argumentative. If you're talking to another ham in the United States, you don't give up any of your First Amendment rights when you communicate via ham radio. I happen to enjoy talking about topics that are controversial, and don't think that someone who disagrees with me is evil or an enemy to be smashed. (In fact, I often learn something from people I disagree with on a subject!) If someone doesn't like what I'm saying, they can tune to another frequency, and I can do the same when I hear something I don't like.

By the way, there's no restrictions on which countries you can communicate with by ham radio. There's no list of "banned" countries you're not supposed to communicate with, nor does the FCC, FBI, CIA, etc., keep tabs on people who talk to other hams in the Russia, Cuba, China, etc. Some nations do ban their hams from communicating with hams in other countries, but that's not your problem if they answer one of your CQ calls.

Basic Equipment in a Ham Station

A call sign, Q signals, phonetics, the RST system, and a knowledge of operating procedures won't be of much value unless you have some way of communicating with other hams. That means you'll need some equipment!

Last chapter, we mentioned that the three basic items are a *transmitter* to send signals to other hams, a *receiver* to receive signals from them, and an *antenna* for the transmitter and receiver. Today, the transmitting and receiving units are usually combined into a single *transceiver*, as shown in Figure 2-4. In fact, at the time this book was being written, no new separate transmitters or matched receiver/transmitter units intended for ham radio use are being offered by ham equipment manufacturers; only transceivers are available. However, there are high performance shortwave receivers (known as *communications receivers*) suitable for ham radio use, and these can be used in conjunction with the transmitting section of a transceiver. Some hams prefer this arrangement since some communications receivers perform better than the receiver section of most transceivers. Just to keep things simple, we'll use the term "transmitter" throughout the rest of this book to indicate either a physically separate transmitter unit or the transmitting circuitry of a transceiver.

FIGURE 2-4: This is a "dream rig" for a lot of hams! The Kenwood TS-850S is a deluxe transceiver whose performance is matched by its price tag.

Many transceivers have a separate *power supply*. As the name suggests, this is a circuit that takes power from the AC line in your house and converts it to the different voltages needed by a transceiver. A power supply could be built into a transceiver (and actually is in a few cases), but is usually separate because the heat from its circuitry could damage or affect the operation of the components making up the transceiver. Line voltages also differ around the world, and it's easier to manufacture one model of transceiver but differing power supply units for each of the countries where the transceiver is sold. (Hey, those Japanese manufacturers aren't dumb!) Many transceivers can operate from 12 volt direct current (abbreviated V and *DC* respectively) power sources, such as those in automobiles. Having an AC power supply built-in would just add weight, size, and circuit complexity.

Antennas for ham stations are more complex affairs than those for your TV set or a CB radio. We'll talk about antennas in detail later in this book, but for now remember there are two broad antenna types: *non-directional* (or *omnidirectional*) and *directional*. A non-directional antenna will radiate power from a transmitter or transceiver equally well in all directions (or in a 360° *radiation pattern*). Such an antenna won't favor stations in any one direction over others from your QTH. By contrast, a directional antenna will have a pattern that concentrates more power in certain directions than others. For example, an antenna could concentrate more power in an easterly direction than to the north, south, or west. Directional antennas are usually designed so they can be rotated to radiate transmitter energy in a desired position. Directional antennas are known informally as "beams."

Non-directional antennas are almost always simpler in design, easier to install, and less expensive than directional antennas. They are also less noticeable than directional antennas, and are thus popular with some hams (such as AA6FW) who live in housing developments with restrictions on antennas. Directional antennas usually are installed on a tower and need a rotor so the radiation pattern can be placed in the direction you want. This all means directional antennas are more complex in design and installation, more expensive, and are considered eyesores by non-ham neighbors and family members. But the increase in performance is tremendous; one of my dreams is one day to have enough land and distant neighbors so I can erect what hams call an "antenna farm"—a collection of large, elaborate directional antennas!

One reason why VHF/UHF has become more popular in recent years involves antennas. The physical size of antenna is a function of the frequency range it is intended to operate on. A good general rule of thumb to remember is this: as the operating frequency increases, the physical size of the antenna needed decreases. This means that a directional antenna for a band like 2 meters (144 to 148 MHz) is no bigger or more elaborate than an ordinary outdoor TV antenna. And you can install an omnidirectional VHF/UHF antenna in situations (such as apartments and condos) where it would be difficult or impossible to install a satisfactory HF antenna.

You'll need a cable to carry the power from your transceiver to your antenna, unless you're using a walkie-talkie type of transceiver where the antenna is connected directly to the transceiver. The cable connecting the transmitter and antenna is known generically as a *feedline*. The most common type of feedline is *coaxial cable*, usually referred to as "coax." Coax consists of a center wire surrounded by a flexible insulating material. This, in turn, is completely surrounded by a braid-like metal "shield." Surrounding this, in turn, is a weatherproof jacket made from rubber, plastic, or other flexible material. Coax is used as the feedline between CB radios and their antennas, and the cables from your VCR to TV are also coaxial.

If you have to tune or adjust your transmitter, there's the strong possibility of causing interference to other licensed hams. A *dummy load* is a device which can harmlessly dissipate the power from a transmitter as heat without radiating it as radio energy.

If you use more than one antenna, or a dummy load and an antenna, you need some way to direct the power from your antenna to the desired antenna or dummy load. This is done with an antenna *switch*. The antenna switch isolates the unwanted antennas (and dummy load) from each other. If you're using separate receiver and transmitter units, you need some way to allow them to use the same antenna. This is done through a *transmit/receive (TR) switch*. A TR switch will "sense" when radio energy is being produced by the transmitter, and will disconnect the receiver from the antenna and instead switch the transmitter to the antenna. When the energy from the transmitter ends, it will be disconnected and the receiver will be reconnected.

The power of almost all ham transceivers is well below the maximum permitted on a given band. To increase power up to the legal limit, an external *radio frequency* (RF) *amplifier* can be used to amplify the output of your transceiver. A special type of amplifier used with some phone modes produces an output signal that varies in exact proportion to the input signal; this type of amplifier is known as a *linear amplifier*. You'll sometimes hear any type of RF amplifier incorrectly referred to as a "linear."

Figures 2-5 and 2-6 show how the basic items of station equipment we've mentioned so far should be connected together. You'll might have a question or two on the written exam based upon diagrams like these. Usually such questions will ask you to identify an item of equipment represented by an unlabeled block in the diagram(s).

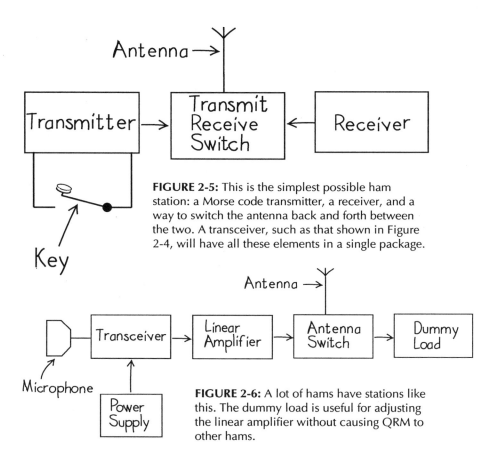

FIGURE 2-5: This is the simplest possible ham station: a Morse code transmitter, a receiver, and a way to switch the antenna back and forth between the two. A transceiver, such as that shown in Figure 2-4, will have all these elements in a single package.

FIGURE 2-6: A lot of hams have stations like this. The dummy load is useful for adjusting the linear amplifier without causing QRM to other hams.

Note the symbols used for the antenna, microphone, and telegraph key in Figures 2-5 and 2-6. These are known as *schematic circuit symbols*, and they are used in circuit diagrams known as *schematic diagrams*. Schematic diagrams are the "road maps" of electronics, and are used to show how the different parts of an electronic device are connected together. For the Novice and Technician exams, you'll have to be able to recognize the most commonly used schematic symbols. We'll take a closer look at schematic symbols in Chapter 6.

There are other items you usually find in a ham shack, and we'll take a look at them in later chapters.

QSL Cards, QSLing, Etc.

One of the oldest traditions in ham radio is for two stations that have been in contact to send QSL cards to each other to confirm that the QSO indeed took place. In the early days of ham radio, hams sent each other QSL cards for every QSO. Today, QSL cards are generally sent out only in special situations—such as DXing and contests—or upon request.

QSL cards range from the simple to the elaborate, but a valid QSL (i.e., one that's acceptable for most operating awards) includes the call sign of the station contacted, the date and time the contact occurred, the frequency the contact took place on, the mode used, and the signature (usually just the first name) of the operator sending the QSL.

To be honest, a lot of hams don't send out QSLs or even have any to send out. The reasons behind this are usually the cost of QSLs and a lack of interest. What's my attitude? If I receive a QSL card, I will answer it. If I want a QSL from somebody, I'll send mine first along with some form of return postage (a self-addressed stamped envelope for U.S. stations, mint stamps or International Reply Coupons—available at U.S. post offices and redeemable at foreign post offices for stamps for postage back to the U.S.—for foreign stations). And, if possible, I'll try to ascertain during our QSO whether the other station will QSL. (Unfortunately, a lot who say they will, won't.)

Given the expense of sending QSLs to foreign stations, a system of *QSL bureaus* have been established by major national ham radio organizations around the world. In the United States, the ARRL operates an incoming/outgoing QSL bureau. The incoming receives and processes QSL cards received from overseas QSL bureaus. To receive cards from the ARRL incoming QSL bureau, all you have to do is keep a supply of self-addressed stamped envelopes on file with the bureau. To send QSLs to foreign hams via the ARRL outgoing bureau, all you have to do is send the cards to ARRL along with a nominal fee ($2 per pound of QSL cards at the time this book was written) for processing.

One of the old sayings of ham radio is "the QSL is the final courtesy of a QSO." A supply of a few hundred QSL cards is available from many specialist printers at a reasonable cost. Even if you're not interested in QSLing yourself, why not keep some cards on hand to reply to those you receive? After all, ham radio is supposed to be fun, and for a lot of hams swapping QSL cards is a big part of that fun.

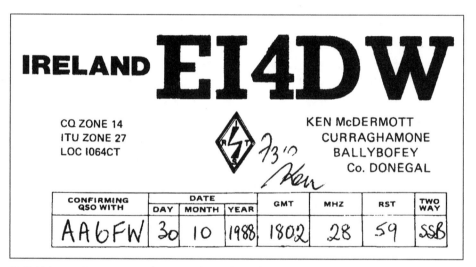

FIGURE 2-7: One reason why many hams look forward to an envelope from the QSL bureau serving their area is the possibility of finding a card inside like this one from EI4DW in Ireland.

Logbooks and Recordkeeping

Until a few years ago, ham radio operators had to keep detailed "logbook" records of all their activities. Hams had to keep a written record of the date, starting and ending times, frequency, transmitter power, mode of operation, and any station contacted for each transmission made. Even unsuccessful CQ calls and tests had to be "logged."

Today you no longer have to keep a logbook of routine contacts and other operating activities. The only records you have to keep are for special or unusual circumstances such as if you operate a repeater station that exceeds certain power limits, remote control operations, and the names and call signs of guest operators who use your station under your call sign. In some cases, such as interference to other radio stations, the FCC can still require you to keep a log of your operating activity.

However, a lot of hams still keep some sort of logbook. This is especially useful if you are into QSLing, DXing, certificate hunting, or some other activity that requires that you have some sort of record of your achievements. I also find a log useful for noting patterns in which areas of the world can be contacted at different times. For example, from my logbooks I know that eastern Europe can often be contacted from southern California late summer nights on 15 meters during periods of high solar activity. I also know that I can usually contact all of Europe easily on 10 meters in the fall around 1600 UTC during periods of moderate to high solar activity. I don't note all my contacts or other operating activities in my log, only "significant" contacts such as DX. Other hams like to make their logbooks into diaries, and include such details as the nicknames of the operators they contact, comments about the performance of their equipment, and related miscellany. The exact form and level of detail is up to you.

Telling Time. . . .

Since ham radio operators can communicate with each other all over the world, they need some way of referring to the time that has some meaning everywhere. Their choice is *coordinated universal time* (abbreviated *UTC*, from the French term for it), which is a fancy new high-tech atomic clock name for *Greenwich mean time* (GMT). This is a time system based upon the 0° meridian at Greenwich, England.

UTC/GMT uses a 24-hour clock system, much like the military does. In UTC/GMT, midnight is given as 0000. The next hour (1:00 a.m. at the Greenwich meridian) is given as 0100. The time 15 minutes later is 0115. The system continues with 0200, 0300, 0400, and so on until UTC afternoon is reached. The next minute following 1259 UTC is 1300 UTC. The time continues with 1400, 1500, 1600, and so on until 2359 UTC is reached; one minute later is 0000 UTC and the start of a new UTC day.

To convert UTC/GMT into your local time, subtract the following number of hours from UTC/GMT:

Time zone	Subtract from UTC for local time
Atlantic standard	Four hours
Atlantic daylight	Three hours
Eastern standard	Five hours
Eastern daylight	Four hours
Central standard	Six hours
Central daylight	Five hours
Mountain standard	Seven hours
Mountain daylight	Six hours
Pacific standard	Eight hours
Pacific daylight	Seven hours
Alaskan standard	Nine hours
Alaskan daylight	Eight hours
Hawaiian standard	Ten hours

An important point to remember is that UTC refers to the day as well as the time. For example, if you have a schedule to contact another ham at 0300 Wednesday, and you live in the Eastern standard time zone, you would try to make contact at 10:00 p.m. *Tuesday* night. Forgetting to make the necessary day conversion in addition to the time conversion is a common error when using UTC.

Taking It on the Road

When you travel outside the United States, you don't have to leave ham radio behind. Many countries have *reciprocal operating* treaties with the United States, in which they agree to honor ham licenses issued by the United States and the

United States returns the favor. These reciprocal privileges are usually obtained by applying in advance for an operating permit from the country you intend to visit. The conditions of the permit and operating regulations are determined by the host country. If you're ever planning to visit a foreign country and want to get in a little operating from there, the ARRL can give you assistance with where and how to apply for a permit. FM on 2 meters is popular around the world, although in much of the world the band is restricted to 144 to 146 MHz. Many users are codeless hams restricted to VHF and UHF, and an American call sign on a repeater in London or Hamburg soon produces a flood of callers!

A happy exception to normal reciprocal operating procedure involves the United States and Canada, which automatically grant reciprocal recognition of each other's ham licenses without the necessity of applying for a permit. For example, you can use a Novice or Technician license in Canada to the full extent of your American privileges. (However, you must observe all Canadian laws and regulations, especially those involving power output and band allocations.) To operate in Canada, just take along your U.S. license when you visit. I've taken along a 2 meter handy-talky on visits to Canada and have really enjoyed chatting with the VE gang from my hotel room!

CHAPTER THREE

Signals, Sidebands, and Other Stuff

YOU TALK INTO THE MICROPHONE of a ham radio station and someone else hears you miles away. Or maybe you're at your computer and dispatch a message via packet radio to a friend on the other side of the country. Regardless of the method you use, you're adding "intelligence" to the signal coming out of a transmitter or the transmitter section of a transceiver. This process is called *modulation*, which means we change that signal in some method (called a *mode*) so that it can convey our voice, picture, or personal computer output to another ham with the proper receiving equipment. (The process of extracting the intelligence from a radio signal is known as *demodulation*.) In this chapter, we'll look at the different ways we can modulate a radio signal and its effects on how you can communicate.

It Starts with the Carrier

Remember what it sounds like when there's some "dead air" between songs on your local radio station? You can tell a signal is there, since the background noise you hear between stations is "quieted," but you hear nothing else. The thing that you're hearing is the *carrier* the station's transmitter is sending out. The carrier is the "raw" unmodulated signal a transmitter puts out. It's called the carrier because the intelligence we add by various modulation methods is "carried" by the carrier to you.

The carrier from a transmitter can be thought of as looking like (if we could actually see it!) part A of Figure 3-1. The carrier is at a constant level of strength, or *amplitude*. Because of this, a carrier is sometimes described as a constant amplitude radio frequency (RF) signal. But how can we alter it in some way so it can convey intelligence?

One way is to simply turn it off and on in a pattern, or *key* it, as shown in part B of Figure 3-1. That's the whole idea behind CW (remember that stands for "continuous wave") telegraphy by Morse code. The carrier's amplitude remains the same, and we haven't altered the frequency or any other parameter in part B of Figure 3-1. Yet the simple act of turning the carrier on and off according to the patterns of the Morse code adds intelligence to the carrier, and—despite its idiot simplicity—is a *bona fide* form of modulation.

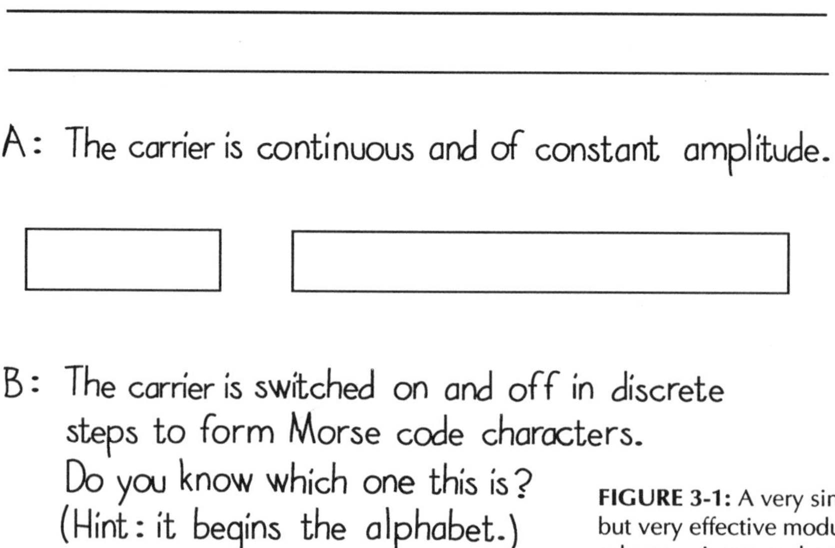

A: The carrier is continuous and of constant amplitude.

B: The carrier is switched on and off in discrete steps to form Morse code characters. Do you know which one this is? (Hint: it begins the alphabet.)

FIGURE 3-1: A very simple but very effective modulation scheme—just turn the transmitter on and off in a pattern.

We can also modulate a carrier by switching it back and forth between two frequency points, as we mentioned last chapter in our discussion of RTTY and other digital modes. Figure 3-2 shows how this works. At part A of Figure 3-2, the carrier is at rest at the mark (higher) frequency. Let's suppose we now start sending Baudot characters via RTTY. When we do so, the carrier drops 170 Hz in frequency, as shown in part B. Note carefully that the carrier isn't interrupted, as with CW telegraphy, and its amplitude remains constant; all that has happened is that the carrier "slides down" in frequency. As we learned last chapter, this lower frequency point is called the space frequency. Part C shows the carrier being shifted back to the mark frequency where it started. Again, all we've done to the carrier signal is shift its frequency upward 170 Hz. This method of modulating a signal by rapidly shifting it between two fixed frequency points is known as *frequency-shift keying* (FSK).

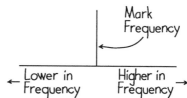

A: The radio signal at rest on the mark frequency.

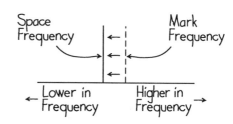

B: The radio signal "drops" to the space frequency.

While effective and simple, both CW telegraphy and FSK are limited modulation methods because all we can do is transmit just two bits of intelligence—dit/dah or mark/space—using them. How do we get a carrier to convey the human voice or

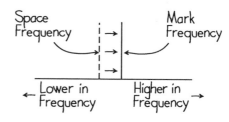

C: The radio signal returns to the mark frequency.

FIGURE 3-2: Frequency shift keying (FSK) illustrated! The carrier stays on all the time, but moves between two frequencies (the mark and space) to form characters.

television images? We'll find out, but first we have to understand the carrier signal better and some ways of describing it.

Testing, testing. . . .

When is an unmodulated carrier not an unmodulated carrier? According to the FCC, when it's a *test* signal.

What's a test signal, according to the FCC? It's an unmodulated carrier of constant amplitude and frequency.

This might seem like splitting hairs, but recent Technician class written exams have often had a question about what a "test signal" is or what you call a signal that does not have any sidebands produced by modulation. The answer is a test signal, a.k.a. a plain old unmodulated carrier.

Sine Waves and Waveforms

To understand more complex modulation methods used to transmit voice or pictures, we have to understand some more about the theory of signal *waveforms*.

The carriers depicted in Figures 3-1 and 3-2 actually consist of a series of sine waves, as shown in Figure 3-3. A sine wave is one that starts from a "zero point" known as the *time line*. The time line is just a handy reference point for the passage of time and the amplitude and position of the waveform at any given moment in time. Starting from the time line, the waveform gradually increases in a positive direction until it reaches a *peak positive amplitude*. When it reaches the maximum positive amplitude, it declines in amplitude until it returns to the time line. The sine wave then starts increasing in amplitude again, but this time in the opposite or negative direction. The amplitude climbs in the negative direction until it reaches a *peak negative amplitude* that is equal to the peak positive amplitude. The wave then declines toward the time line until it returns to the time line. When a sine wave has made a complete trip from the time line to the peak positive amplitude, back to the time line, then to the peak negative amplitude, and finally back to the time line, it is said to have gone through one complete *cycle*.

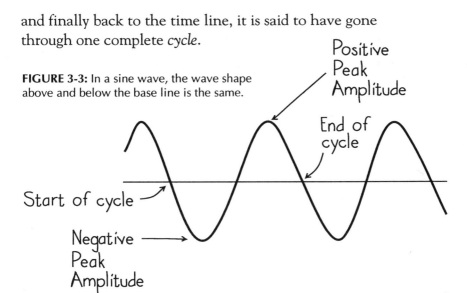

FIGURE 3-3: In a sine wave, the wave shape above and below the base line is the same.

When you think about it, a cycle is a bit like a circle—it winds up back where it started from (that is, the time line). A complete sine wave cycle, like a circle, is divided into 360 degrees and we refer to where a sine wave is at during a cycle by degrees. Figure 3-4 shows how a sine wave is divided into degrees. A sine wave starts at 0°, has its peak positive amplitude at 90°, returns to the time line at 180°, reaches a peak negative amplitude at 270°, and returns to the time line at 360°, which happens to be the 0° position for the next sine wave cycle. The "position" of a sine wave during a cycle is called its *phase*. When a sine wave is at its peak positive amplitude, its phase is 90°.

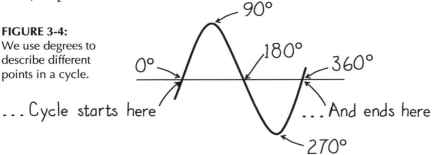

FIGURE 3-4: We use degrees to describe different points in a cycle.

In chapter one, we saw how frequencies are measured in hertz (Hz), kilohertz (kHz), and megahertz (MHz). "Hertz" is a measure of how many cycles of a sine wave take place within one second. For example, the AC (alternating current) electricity flowing into your home is rated at 60 Hz, meaning there are 60 sine waves of the AC electricity in a second. As we increase in frequency from Hz to kHz to MHz, more cycles are "crammed" into each second. The frequency of 3900 kHz—or 3.9 MHz—in the 75-meter ham band means that a signal of that frequency goes through 3,900,000 cycles in a second. Figure 3-5 shows low frequency and high frequency sine waves. Both reach the same peak positive and negative amplitudes, but there's a clear difference in the number of cycles that occur in a period of time.

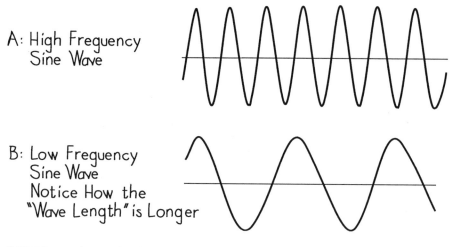

A: High Frequency Sine Wave

B: Low Frequency Sine Wave Notice How the "Wave Length" is Longer

FIGURE 3-5: The amplitude (or height, meaning the strength) of these two sine waves is identical. However, many more cycles happen at A in a period of time than do at B. This means A has a higher frequency.

A sine wave is an example of a waveform, which is how a signal would look if we could see it. There's a common electronic instrument known as an *oscilloscope* that lets us do just that. Think of an oscilloscope as a television set that lets you see waveforms. A signal (such as that from a radio transmitter) is applied to the inputs of an oscilloscope, and the waveform is displayed on the oscilloscope screen. (You might wonder how you can see a signal with thousands or millions of cycles taking place in just a second. Because of the regularity of a sine wave, the displayed waveform appears to be "standing still" even though it's constantly changing.)

Sine waves are not limited to radio signals or AC electricity. A pure, steady audio tone is also a sine wave. You may have heard a test record or tape, such as those used to test audio equipment, which has various audio tones of fixed frequencies. You might also be familiar with a device known as an audio signal generator often found in electronic labs to generate audio signals of a precise frequency. If you were to apply an audio frequency from one of these sources to an oscilloscope, you'd get a sine waveform displayed.

Much of the information we transmit by radio is not in the form of a sine wave, however. The human voice is a good example. When you speak, you don't speak in pure sine wave tones. Instead, your voice is a mixture of sounds of varying frequencies and amplitudes. If you speak into a microphone and display the resulting electrical signal on an oscilloscope, you'll get something like Figure 3-6—the signal will vary in both frequency and amplitude over a wide range and in no discernable pattern. (If anything, the waveform in Figure 3-6 is overly simplified.) Our task in trying to modulate a transmitter is to figure out some way to add that chaotic audio or video waveform to the carrier signal.

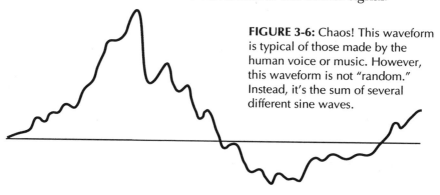

FIGURE 3-6: Chaos! This waveform is typical of those made by the human voice or music. However, this waveform is not "random." Instead, it's the sum of several different sine waves.

Bandwidth

We also need to understand the concept of *bandwidth* before looking at specific modulation methods. Bandwidth means the amount of frequency space a radio signal occupies. This amount varies with the modulation method used. As a general rule, the more "intelligence" there is in a radio signal, the more bandwidth the signal occupies.

An unmodulated carrier occupies very little frequency space, but if we send Morse code by interrupting the carrier, we slightly increase the bandwidth of the signal. The usual rule of thumb is that the bandwidth of a CW signal can be found by multiplying the speed in words per minute (WPM) by 4. A signal of 20 WPM will occupy about 80 Hz, while a 30 WPM CW signal will occupy 120 Hz of bandwidth. The extra bandwidth occupied by the faster code speed reflects the additional amount of intelligence present in the faster code speed (that is, more information is being transmitted in a given amount of time). The same thing happens with RTTY and digital signals; the faster information is transmitted (the higher the "baud rate"), the more bandwidth such signals will occupy.

By now you might suspect that adding sound or pictures onto a RF carrier will cause the bandwidth of the signal to really explode. If you do, you're absolutely right. The narrowest bandwidth possible for a voice signal is a few kHz, while television takes up a few MHz!

Amplitude Modulation (AM)

This is the oldest form of voice modulation, and many hams jokingly refer to it as "ancient modulation." It's still used today on the AM broadcast band (540 to 1600 kHz), and is also used to transmit the video (but not sound) portion of television signals. In a nutshell, amplitude modulation is when we make the amplitude (power level) of a RF signal from a transmitter vary in direct proportion to the variations of an audio signal (voice or, in the case of commercial broadcasters, music) used to modulate it. At the peak positive amplitude of the modulating signal, the output of the transmitter is at maximum; at the peak negative amplitude of the modulating signal, the output of the transmitter is at a minimum.

In amplitude modulation, there is always a carrier present (for example, you can always tell a carrier is present on an AM broadcast station during "dead air" between songs and announcements). To convey intelligence, we take the electrical signal from a microphone, amplify it greatly in strength, and *add* it to the carrier—during amplitude modulation, we're adding power to the output of the transmitter! The result is an output signal consisting of the carrier and *sidebands* (or *sideband frequencies*) above and below the carrier. The sidebands are "mirror images" of each other, and each is equal in bandwidth to the highest audio frequency used to modulate the transmitter. For example, if we were to modulate an AM transmitter with a 5 kHz audio sine wave, the sidebands would extend to 5 kHz above and below the carrier. The carrier does not add any intelligence to the signal, but instead (like the word implies) "carries" the sidebands. Since the sidebands are identical, the total bandwidth of an AM signal is equal to *twice* the highest audio frequency used for modulation. To continue with our example, an AM signal modulated by a 5 kHz sine wave would have a total bandwidth of 10 kHz.

Figure 3-7 shows how this process works. To keep things simple, let's assume the modulating audio signal is a sine wave as shown in part A of Figure 3-7. The unmodulated carrier from the transmitter will be at a constant amplitude, as seen in part B. But when we amplify the modulating signal and add it to the carrier in the transmitter, the result is what you see in part C of Figure 3-7. The energy in the output signal above the level of the unmodulated carrier is the result of the amplified audio signal, and the sidebands extend out above and below the carrier according to the frequency of the modulating audio signal. The waveform you see in part C is the result of adding the carrier and two sidebands together; if you were to separate them, you'd have three sine waveforms.

A: Sine Wave
   Modulating Signal

B: Unmodulated
   Carrier from
   Transmitter

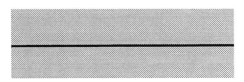

C: The Modulated Signal!
   Note How the
   Amplitude is Greater
   than that of the
   Unmodulated Carrier

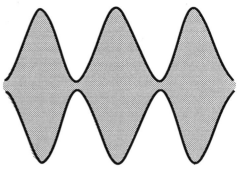

FIGURE 3-7: Amplitude modulation is the process where additional power is added to the output of a transmitter in accordance with variations in the modulating signal.

Another way to visualize an AM signal is the by-frequency graph shown in Figure 3-8. The two sidebands are referred to by comparison to the carrier. The sideband lower in frequency than the carrier is called the *lower sideband* (LSB) while the sideband higher in frequency than the carrier is the *upper sideband* (USB). The frequency of an AM signal is measured by the frequency of the carrier. For example, if you're listening to an AM broadcasting station operating on 1100 kHz, then 1100 kHz is the carrier frequency and the sidebands lie above and below that frequency.

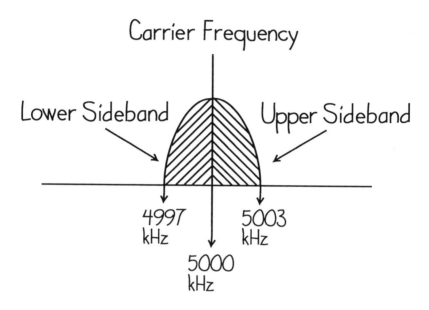

FIGURE 3-8: The result of amplitude modulation is two sidebands located above and below the carrier.

The more modulation we can add to a carrier, the louder an AM signal will sound—up to a point. That point is when the output signal rises to twice the level of the unmodulated carrier at the positive peak amplitude (or 100%) of the modulating signal. If we try adding more of the modulating signal and boost the output to over twice the unmodulated carrier level, we *overmodulate* the output signal. When this happens, the signal becomes distorted and hard to understand. Even worse, something called *splatter* happens. An overmodulated AM signal produces more than just two sidebands; it also produces a slew of distorted, spurious signals above and below the carrier, often many kilohertz away. These signals interfere with other stations on adjacent frequencies. To prevent this, the FCC requires all ham stations to limit their modulation to no more than 100% and also take any additional steps necessary to prevent splatter or other spurious signals.

AM transmitting and receiving circuits are simple, and it's easy to tune in an AM signal. But AM has a couple of big disadvantages. For one thing, it wastes frequency space since an AM signal takes up twice as much frequency space as the highest audio frequency transmitted. Perhaps even more importantly, AM is an inefficient mode. Two-thirds of the power of a 100% modulated AM signal is in the carrier, which conveys no information. The other one-third is in the sidebands. However, since the sidebands are mirror images of each other and convey the same information, one of the sidebands is unnecessary and wasteful. So, of the the total output of an AM transmitter, only one-sixth actually conveys intelligence.

You might be thinking that it would be smart if we could just transmit one sideband instead of two sidebands and a carrier, putting all of the transmitter power where it would do the most good. Some clever hams thought of that many years ago, as we'll see in our next section.

What's the Audio Limit???

Something may have occurred to you while reading the discussion about the bandwidth of an AM signal. If each sideband is equal to the maximum audio frequency used to modulate the carrier, then a 3 kHz tone will produce an AM signal with a 6 kHz bandwidth. If a 5 kHz tone is used, the result will be a signal with a 10 kHz bandwidth. As you kept using higher audio frequencies to modulate the transmitter, you would increase the bandwidth of the transmitted signal. And this same general principle also holds true for frequency modulation (FM) signals: the higher the audio frequency transmitted, the more bandwidth a signal occupies.

However, we don't need to transmit such high audio frequencies for ham radio communications. High audio frequencies are important for transmitting music with high fidelity, but we hams can't transmit music. It so happens that all the audio frequencies necessary for clear, intelligible transmission of the human voice lie at audio frequencies below 3 kHz, and this is generally the maximum audio frequency transmitted by hams. Several ham transceivers, in fact, are designed to be unable to transmit frequencies above 3 kHz!

Single Sideband (SSB)

Single sideband is a modulation mode in which one sideband and the carrier are suppressed and only one sideband is transmitted over the air. The theory behind SSB was developed in the 1920s, but it wasn't until after World War II that hams began to experiment with it over the air. Hams were among the very first to use this mode, and today SSB is the most widely used voice mode by ham, military, aeronautical, and shipboard stations on frequencies from 1.6 to 30 MHz.

In most SSB transmitters, an AM signal is produced at a low power level. The carrier and one sideband are then removed, and the remaining sideband is then greatly amplified by the transmitter circuitry. How is the carrier and unwanted sideband removed? A *filter* does this job. We'll discuss filters in more detail later, but for now just remember that a filter is a device that lets a certain bandwidth of frequencies pass but rejects all other frequencies outside that bandwidth. In SSB transmitters, the filter is just wide enough to let one sideband pass. The other sideband and carrier are blocked by the filter. An AM signal with two sidebands and a carrier enters the filter; just one sideband comes out. The remaining sideband is then boosted in power by an amplifier circuit before it's transmitted.

It really makes no difference which sideband—LSB or USB—we transmit, since both are identical. However, we have to tune our receivers a little differently depending on whether we want to receive LSB or USB. To prevent confusion, hams normally transmit LSB on 160, 80, and 40 meters and transmit USB on 20 meters and higher. This is just normal operating practice, though, and is not required by the FCC.

How much better is SSB than AM? For one thing, you can squeeze two SSB signals in the space occupied by one AM signal. The increase in communications efficiency is startling. How much more effective depends on whose mathematical theorems you care to believe, but as a good general rule a SSB transmitter is four times more effective than an AM transmitter of the same power!

If SSB is so terrific, how come the AM broadcast band from 540 to 1600 kHz hasn't changed over to the "SSB broadcast band"? The main reason is that the carrier doesn't do much good when a signal is transmitted, but it's a different story with reception. The carrier is an essential part of extracting intelligence from the signal—called *detection* or *demodulation*—and the lack of a carrier means a SSB signal is unintelligible when received on an AM receiver. SSB on an AM receiver is hard to describe; perhaps the best way is to imagine what Donald Duck would sound like after inhaling helium! In ham SSB transceivers, this problem is solved by generating a replacement carrier in the receiving circuitry. However, this complicates tuning too much for average consumers, so SSB remains off-limits for commercial broadcasting. But for hams, SSB is by far the most efficient voice mode around. It "gets through" under conditions (weak signals, heavy interference, etc.) where other phone modes can't.

By the way, the same warnings about overmodulation mentioned in the previous section on AM also apply to SSB. Most SSB transceivers have a control labeled "microphone gain" or similar. Some hams crank this all the way up trying to make their signal sound louder. Unfortunately, beyond a certain point increasing this control will just cause overmodulation and interference to nearby frequencies. Some SSB transceivers also have what are known as *speech processing circuits*. These circuits try to keep the amplitude of a SSB transmitter's output as constant as possible for highest possible average power. If set too high, these can produce distortion and spurious signals on adjacent frequencies. While speech processing circuits can make a SSB signal sound louder, they often make the audio of the signal sound "artificial" and unpleasant. As a result, many hams (like me) don't use a speech processing circuit unless absolutely necessary because of heavy interference or poor conditions.

Frequency Modulation (FM)

There are other ways to add intelligence to a carrier other than varying its amplitude. We can let the amplitude of the carrier remain constant, and instead vary it above and below its unmodulated frequency according to variations in the modulating signal. This is frequency modulation, and it's the most popular mode for phone communication on frequencies above 50 MHz. The same advantages that have made the FM broadcasting band so popular for music listeners—the excellent fidelity and freedom from noise—also apply to FM in ham radio. Since FM is so important in ham radio today, you can expect to see several questions about it on the written exam for your license.

Figure 3-9 is a really simplified look at how FM works. Part A of Figure 3-9 shows the unmodulated carrier from an FM transceiver. This is the signal you would produce if you were to press the microphone button of an FM transmitter but not say anything. When you say something into the microphone, as shown in part B, the frequency of the carrier "swings" upward. This swing in frequency due to the modulating signal is known as *deviation*. The carrier swings above and below its unmodulated frequency, and the unmodulated frequency is known as the *center frequency*. The maximum amount of deviation from the center frequency is set by the FM transmitter circuitry, and for ham radio communications this is 5 kHz. Since a signal will deviate above and below the center frequency, the total bandwidth for FM phone in ham radio is 10 kHz. This same 5 kHz standard is used by police, fire, business, and other two-way radio services you can hear on a scanner. The greater the deviation, the wider the range of audio frequencies that can be transmitted by FM. Five kHz is plenty of deviation for the human voice, but FM broadcast stations use deviation of 75 kHz (!!) to transmit music.

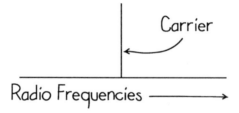

A : Unmodulated FM carrier "at rest" on the center frequency.

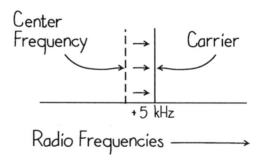

B : Modulation "deviates" the carrier 5 kHz higher in frequency.

FIGURE 3-9: The process of frequency modulation is similar to that for FSK—the carrier remains on constantly and at the same amplitude but varies from the unmodulated frequency. Unlike FSK, the carrier of an FM signal varies over a continuous range instead of being switched between two fixed frequency points.

An important point to remember about FM is that the amount of deviation is proportional to the *amplitude* of the modulating audio signal, not its frequency. This is sometimes a hard point to grasp—it's a bit contrary to our common sense expectation that a higher audio frequency should shift the carrier frequency higher—but it's true. The louder you speak into the microphone of an FM transmitter, the more deviation you cause. Figure 3-10 shows how this works. When no modulating signal is applied to an FM transmitter, the carrier remains at its center frequency. But as the amplitude of the modulating signal rises to its positive peak, the carrier frequency deviates *upward*. In Figure 3-10, notice how the cycles "compress" at the positive peak; this compression means there are more cycles per second—the frequency is higher. But you can also see that the amplitude of the carrier hasn't changed a bit. As the amplitude of the modulating signal drops down to its negative peak, you can see that the carrier frequency also drops down, eventually dropping below the center frequency and reaching a maximum deviation below the center frequency at the negative peak of the modulating signal. The carrier amplitude still remains constant, even though its frequency is constantly "on the move" around the center frequency.

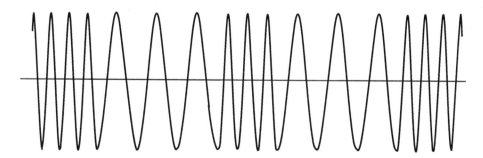

FIGURE 3-10: The waveform of an FM signal. The carrier amplitude remains constant, but the frequency varies according to the modulating signal.

Like AM, FM produces sidebands as a result of the modulation process. Unlike AM, FM produces multiple sets of sidebands above and below the center frequency. The number of sidebands generated by FM depend on the frequency of the modulating signal and the deviation used. And it's also possible to cause interference to stations on adjacent frequencies with an incorrectly operated FM transmitter. If too much modulating signal is applied to the transmitter, *overdeviation* results. The results are the same as when an AM transmitter is overmodulated—spurious signals are found outside the normal signal bandwidth (10 kHz in this case) and cause interference. One solution to overdeviation is to simply speak more softly into the microphone, since the deviation is proportional to the strength of the modulating signal. Some FM transmitter microphones have an adjustable "gain" control, which determines how strong an audio signal is delivered from the microphone to the transmitter. In this case, the gain can be lowered. Finally, most FM transmitters have an internal deviation adjustment control. If you're consistently overdeviating, reducing the setting of this control will usually clear up the problem.

Since FM takes up so much more bandwidth than AM or SSB, it's easy to understand why it's restricted to the roomier VHF/UHF ham bands. But SSB is more effective than an FM signal of the same transmitter power, so why use FM at all? The reason is that for local communications FM has some powerful advantages over other modes. One is its high immunity to noise. It just so happens that most types of electrical noise (from motors, automobile electrical systems, power lines, etc.) and static that interfere with radio communications are amplitude modulated. FM receivers are designed to screen out amplitude modulated signals and to respond only to FM signals, so FM communications are largely free of the noise that can severely degrade SSB or AM signals. Another is that above a certain signal level, the background noise in an FM receiver is quieted and only the received signal is heard (you've probably noticed this when you tune FM broadcast stations). In SSB and AM communications, some background noise remains unless the signal level is extremely high. Finally, there's something known as the *capture effect*. If two or more FM stations are on the same frequency, and one is somewhat stronger than the others, the strongest signal will "capture" a receiver and be the only one heard. These factors mean that the audio quality and communications reliability of FM is far superior to AM or SSB if the strength of a received signal is greater than the background noise in the receiver, which it normally is during local communications. Most FM receivers include a *squelch* circuit, which keeps the receiver quiet until a received signal exceeds a certain level. When a signal at or above that level is received, the squelch "unlocks" and the signal is heard. If a signal is strong enough to unlock the squelch, it is usually well above the receiver background noise. These factors mean that local FM communications through repeater stations is a lot like talking over an intercom—no noise, no interference!

FM's First Cousin—Phase Modulation (PM)

A lot of FM transceivers used by hams aren't really FM— they're *phase modulation* (PM) rigs. This is a different modulation technique, although the end result is about the same. Any receiver that can receive FM can also receive PM, and you really can't tell much difference between the two. (Some hams claim that they can hear a difference between FM and PM, but I can't.)

Phase modulation means, appropriately enough, that we use the audio modulating signal to alter the phase of the carrier. The amplitude of the carrier isn't varied, only the amount of time it takes the carrier to reach certain positions in the sine wave cycle. By varying the phase of the carrier waveform, you also vary the number of cycles that take place per second. And if you do that, you. . . . sonofagun!!! You cause the frequency of the carrier to change—you're really frequency modulating the carrier. And that's why FM and PM are really two different ways of getting the same result. You can't modulate the phase of a carrier without altering its frequency, and you can't modulate the frequency of a carrier without changing its phase. This is why PM is sometimes called *indirect FM*.

There is one key difference between FM and PM you should remember, though. In FM, the change in the carrier frequency depends only on the amplitude of the modulating signal. With PM, the frequency shift depends on both the amplitude *and frequency* of the modulating signal. This explains why some hams claim they can hear a difference between FM and PM; the same modulating signal will produce somewhat different amounts of carrier frequency shift.

So why are some ham VHF/UHF transceivers FM while others are PM? Because a PM transmitter circuit is often a simpler design than a comparable FM transmitter.

FM Repeaters

As mentioned in our first chapter, FM is really popular on the VHF and UHF bands when used in conjunction with repeater stations. Figure 3-11 shows a block diagram of the major elements of a repeater station.

There are a couple of items in figure 3-11 which are important in repeater operation. The *carrier operated relay* (COR) is used to turn the repeater's transmitter on whenever a signal is received on the repeater's input frequency. The "carrier" refers to a radio signal, and the COR keeps the repeater transmitter on as long as there's a signal on its input frequency. When there's no longer a signal on the input frequency, the COR turns off the transmitter. In addition, the COR is usually the place where the repeater's *timer* is located. Once a signal is received on the input frequency and the COR turns on the transmitter, the timer starts running. If the signal on the input frequency ends, the timer resets to zero. But if the input signal continues, the timer will shut off the transmitter after a fixed period of time. This period can be as short as 30 seconds or as long as three minutes; most repeaters allow a full three minutes. The purpose of a timer is to prevent one station from monopolizing a repeater and encourage natural, back-and-forth conversations instead of monologues. The timer also shuts off the repeater in case a malfunction turns the transmitter on when no input signal is being received.

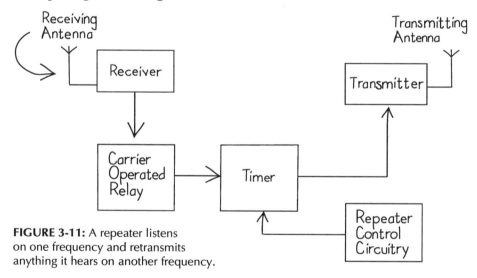

FIGURE 3-11: A repeater listens on one frequency and retransmits anything it hears on another frequency.

A lot of repeaters use separate antennas for receiving and transmitting. However, it's possible to simultaneously use one antenna for both purposes by using a *duplexer*. This is a clever device that electrically "isolates" the receiver and transmitter from each other and allow them to share the same antenna. Duplexers are expensive and complex beasts, however, and separate antennas are often the cheapest solution.

Using FM for CW, RTTY and Packet

The advantages of FM (and PM) are not restricted to phone communications. We can also use FM to send Morse code, RTTY, packet, and other digital modes on the VHF/UHF bands with a high degree of reliability. This is done using audio tones transmitted via FM, much like audio tones are used to link computers over telephone lines.

If you've listened to any ham repeaters on a scanner radio, you've probably heard Morse code sent as audio tones used to identify the repeater while it's in use (as required by FCC rules). Some other repeaters conduct on-the-air Morse code practice using audio tones to form the different characters. This is known as *modulated CW* (MCW). We can also transmit RTTY and packet via FM by switching between two fixed audio tones, much like switching between two fixed frequencies. This is known as *audio frequency-shift keying* (AFSK), and is used to RTTY and packet on the VHF/UHF bands. We'll take a closer look at packet in the next chapter.

Harmonics

A *harmonic* of a signal is an integer multiple of it. For example, a signal at 5 MHz would have harmonics at 10, 15, 20, and 25 MHz. The 5 MHz signal is known as the *fundamental* frequency. 10 MHz is known as the *second harmonic*, 15 MHz as the *third harmonic*, 20 MHz the *fourth harmonic*, and so on. While the frequency will be different, the harmonics will have the same waveform characteristics as the fundamental frequency.

All radio transmitters produce some harmonics, although in a well-designed and properly operated transmitter they're too weak to be of any importance. Are harmonics good or bad? It depends. Sometimes we deliberately want a circuit to have a strong harmonic output, but usually a transmitter that produces strong harmonics is big trouble.

A circuit that deliberately produces harmonics of a signal is known as a *frequency multiplier*. These circuits are found in almost all ham FM transceivers, since it's easier to frequency or phase modulate a lower frequency signal and then "multiply" it up to the desired output frequency. Another common circuit in VHF/UHF work is known as a *tripler*. A tripler takes an input signal at 144 MHz (the 2 meter ham band) and multiplies it to 432 MHz (the 70 cm ham band). In essence, the tripler takes the 144 MHz and deliberately generates its third harmonic. This is an easy way to cover the two most popular VHF/UHF bands with a single transmitter.

However, harmonics can cause interference to other radio services. One common problem involves transmitters operating above 28 MHz on the 10-meter ham band. Multiply that frequency by two, and you get a signal at 56 MHz, which is smack in the middle of television channel 2! It's also possible for a CW transmitter on 7050 kHz in the 40-meter band to radiate strong harmonics at 14100 kHz in the 20-meter band and 21150 kHz in the 15-meter band.

Harmonics are not as big a problem as they were a few years ago, thanks to the solid-state (i.e., no vacuum tubes) transmitting gear used today. The old vacuum tube transmitters had to be carefully tuned to the transmitting frequency being used, and incorrect adjustment often resulted in strong harmonics being radiated. Today's solid-state rigs require no such adjustment, and harmonics are usually not a problem unless an external tube-type power amplifier is being used. (Vacuum tubes still have some advantages over transistors for amplifying transmitter output to higher power levels, and tubes are still widely used for this purpose.) But you should be aware of the possibility of harmonic radiation, since FCC rules require hams to suppress the harmonics produced by their transmitters to the lowest possible level. As we'll see in a future chapter, there are various types of filters which can be placed between your transmitter and antenna to suppress any harmonics that your transmitter might be generating.

Clicks, Chirps, and Hum

Overmodulation isn't the only cause of spurious or distorted signals on the ham bands.

Hum is just what you might think it is—a constant, steady humming or buzzing sound superimposed on a signal in addition to the intended modulation. The hum can be heard in the background of such phone modes as SSB and FM, and as a "rippling" on CW and RTTY signals. If you ever hear a signal suffering from hum, you'll notice how the hum seems to sound like the 60 Hz humming sound produced by large power transformers and AC power lines. This is because hum is the result of a problem in the power supply that converts 120 volt AC power into the different voltages required by the transmitter or transceiver. In the power supply, components known as *capacitors* are used, and sometimes these fail. When they do, the 60 Hz AC power line signal is not adequately filtered out and can sneak into the modulating signal. The result is that a 60 Hz audio tone can be heard on the modulated transmitter signal. The solution to this problem is to replace the defective capacitors. (Yeah, we haven't gotten around to explaining what a capacitor is. . . . we'll cover that soon, so just file this bit of information away until then.)

Chirp is also caused by power supply problems, and is most commonly a problem with CW signals. A "chirp" is a shift in the transmitter's frequency when it is keyed to form Morse code characters. This shift causes the dits and dahs to sound like bird chirps or have a "wobble" in them. The sound is quite distinctive if you ever run across such a signal! Chirp is caused by a power supply that's unable to supply steady voltages to the transmitter while it's being keyed. The varying voltages cause slight frequency shifts with each dit and dah transmitted. The solution is a power supply that has a more stable output voltage, or better *regulation*, when used to power a transmitter.

Key clicks are another problem with CW transmitters. These are similar to the spurious signals produced by overmodulated phone signals, and sound like clicking noises. Key clicks can unfortunately be heard several kHz away from the signal producing them. They are caused by turning the carrier on and off too sharply when forming the Morse characters, which generates spurious signals. Key clicks can be minimized by a *key click filter*, which "softens" the edges of the dits and dahs. Almost all commercially available CW transmitters and transceivers have key click filters built-in, so key clicks are seldom a problem today unless you're using older or home-built equipment.

Emission Designators and FCC Emission Terms

The International Telecommunications Union (ITU) is an international organization that most countries of the world (including the United States) belong to. The ITU divvies up the radio spectrum worldwide and establishes standards for its use. One of the things the ITU has tried to do is develop a standard, precise way to refer to different types of emissions. This system is language independent, and uses two letters and a single digit in a letter–digit–letter format. The first letter denotes the type of main carrier modulation used. The digit indicates the nature of the signals modulating the main carrier, and the last letter denotes the type of information to be transmitted. In years past, many ham exams carried a question or two about these ITU designators. The current ham exams don't require you to know these designators, but they do show up in a lot of ham magazines, equipment owner's manuals, and other places, so you might as well be familiar with the most common ones. Table 3–1 gives these.

TABLE 3–1

Commonly Used ITU Emission Designators

N0N	Unmodulated carrier
A1A	Continuous wave Morse code telegraphy
A2A	Morse code sent using audio tones over an AM signal
A3E	Amplitude modulated voice
A3F	Amplitude modulated television
F1B	Frequency-shift keying
F2D	Frequency modulated data
F3E	Frequency modulated voice
J3E	Single sideband voice

The FCC now uses nine different emission terms in its rules and publications. While you won't be tested on these either (at least under the exams in use when this book was written), you need to know what these terms mean to fully understand the FCC rules and regulations pertaining to ham radio. Here's a brief summary of them:

- CW. International Morse code telegraphy emissions.
- *Data.* Telemetry, remote control, and computer communications.
- *Image.* Facsimile and television signals.
- MCW. Morse code sent using audio tones over a phone transmitter.
- *Phone.* Speech and other incidental sounds.
- *Pulse.* Not currently used extensively in ham radio, this emission is used for advanced computer communications and digital voice signals.

- *RTTY.* Signals to print text at a receiver.
- *Spread spectrum (SS).* Not currently used extensively in ham radio, this is a wide bandwidth modulation technique.
- *Test.* Signals containing no information whatsoever, such as an unmodulated carrier.

Strange Ham FM Lingo. . . .

When FM and repeaters first became used in ham radio during the late 1960s, a whole new set of operating procedures sprang up because of the profound differences between the then-new mode and conventional ham radio operating. These differences included the use of "squelched" receivers, fixed frequency "channels," and repeater stations. The result was somewhat like CB radio, with many people listening continuously to a relatively small number of frequencies. While FM has become more integrated into the mainstream of ham radio, it still has a distinct nomenclature all its own!

band plan: a plan to allocate input and output frequencies for repeater use and simplex frequencies on a standard basis on a national and regional basis. The ARRL has done some great work for hams by sponsoring national band plans (with some regional exceptions) for repeater frequencies and for the VHF/UHF bands as a whole.

closed repeater: a repeater that is for the use only of a certain group of hams, such as a club. Access to the repeater usually requires a certain tone be transmitted at the beginning of a transmission.

channel: the input and output frequency pair used by a repeater, usually identified by the output frequency. For example, suppose a repeater has an input frequency of 146.16 MHz and an output frequency of 146.76 MHz. This "channel" would be referred to as "seven six."

coordinator: a club or other group that helps voluntarily assign frequencies and coverage areas for repeater stations so as to keep interference between repeaters to a minimum.

destinated: to arrive at one's location when operating FM in a car, as in "I've got to sign with you now because I'm destinated." (Often frowned upon.)

direct: to communicate via simplex, often pronounced "die-rect." (This pronunciation often frowned upon.)

full quieting: an FM signal loud enough to completely quiet the receiver background noise.

intermod: short for "intermodulation," this usually means conditions where spurious signals can be heard on the output frequency of a repeater.

kerchunker: someone who activates the repeater by pressing the microphone button but doesn't say anything, causing receivers tuned to the repeater to hear a "kerchunk" sound when the squelch is broken (these people are frowned upon).

machine: a repeater station, as in "I hear they're putting a new machine atop Mt. Baldy."

open repeater: a repeater that can be used by anybody; a signal on its input frequency will automatically be relayed on the output frequency.

picket fencing: a "fluttering" of the signal from an FM station in motion, usually encountered from stations in moving cars. This happens because trees, other cars, and similar objects momentarily block or reduce the signal.

squelch tail: the brief bit of noise heard between the end of a repeated transmission and the activation of the receiver squelch circuit.

split channel: a repeater input/output frequency pair that doesn't conform to the national plan developed by the ARRL, usually by squeezing the pair in between standard channel pairs.

time-out: to talk too long in a single transmission and cause the repeater's automatic timer to shut the machine off.

tone access: a repeater that requires a tone to be transmitted at the start of each transmission in order for the signal to be relayed.

uncoordinated repeater: a repeater that has not coordinated with other repeaters through the auspices of a coordinator. The key word here is "uncoordinated." Like the word implies, these repeaters tend to trip over themselves and others, flop about aimlessly, and generally make life miserable for everybody but their egocentric owners/operators.

CHAPTER FOUR

Ham Radio Meets the Personal Computer

I N THE PREVIOUS CHAPTERS, I've been dropping big hints about how terrific the marriage is between ham radio and personal computers. You might already be communicating over your personal computer (PC) by using a modem and telephone lines. Packet radio—linking computers via ham radio—is a lot like an over-the-air version of popular computer networks such as CompuServe, UseNet, and GEnie. One major advantage of packet is that it allows you to "address" a message to a specific station, which can be routed and delivered to that station even if the other operator is not present at their station when your message arrives. Packet is also the most error-free method of ham radio communication, and can be used to send audio and image data as well as text characters. This is where the action is today in ham radio—and it gets hotter with each passing month—so let's see what all the excitement is about.

It All Starts with a Tink

The box that interfaces your PC to your ham radio transceiver is called a *terminal node controller* (TNC), often called a "tink." This is a radio version of a modem, converting the digital signals from your PC into audio signals that your transceiver can transmit. On receive, the TNC takes audio from the transceiver and converts it into the digital form your computer understands. These processes are controlled by software. TNCs have software in read-only memory (ROM) to control the signal conversion process, but the PC needs some sort of communications software to "talk" to the TNC. Most TNCs are supplied with software for the PC designed for ham radio communications. Figure 4-1 shows a popular TNC that's typical of those used by hams.

FIGURE 4-1: The Pakratt 232 multimode TNC by Advanced Electronic Applications, Inc. gives packet, CW, RTTY, AMTOR, and ASCII capabilities in one package.

There are two general types of TNCs in use today. The simplest and least expensive types are *packet-only* TNCs. Such TNCs provide full packet radio capabilities, but do not support other modes. If you use a packet-only TNC, the only software your PC needs is an ordinary communications or terminal emulation program such as the one you use to communicate via modem over telephone lines. Packet-only TNCs are usually compact and can often be battery powered, making them useful for portable and emergency use. The second type of TNC, the *universal terminal unit* or *multimode terminal,* is now the most popular. These units offer full packet radio capabilities as well as the ability to receive and transmit other modes such as RTTY, pure ASCII text, and even Morse code. Most multimode TNCs are supplied with special communications software to take advantage of all their features, although you can get by using an ordinary terminal emulation program. Both types of TNCs are placed between your PC and ham radio transceiver as shown in Figure 4-2.

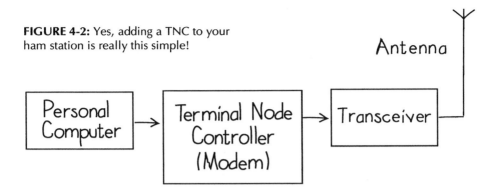

FIGURE 4-2: Yes, adding a TNC to your ham station is really this simple!

Communications between your PC and TNC are usually via a standard RS-232 serial port like that used for modem communications. The connections between a TNC and ham radio transceiver are more complex, and vary widely according to the transceiver type. Most connections are made using the microphone input jack on the transceiver, and a separate microphone plug and cable. One end of the cable is connected to the TNC, while the microphone plug is wired to the other end of the cable. The exact wiring scheme for the microphone plug will depend on the transceiver. Some way is necessary to get audio output from the transceiver for the TNC. Some transceivers anticipate this need and have a separate audio output signal available at their microphone connection jacks (this signal is not used when a conventional microphone is connected). Otherwise, the audio signal can be taken from the transceiver's headphone jack or speaker terminals.

All TNCs let you configure your packet radio station so that it serves as a digipeater for other packet signals. Advanced TNCs and PC software also let you set up an on-the-air BBS and a "mailbox" for packets addressed to you while you're away from your station. We'll discuss such features later in this chapter.

Other Digital Modes with a TNC

We took a look at other digital modes back in Chapter Two. Before looking at packet radio in depth, let's first discuss some of the other modes available on a multimode TNC and its supporting software:

Morse code—Yes, the oldest mode of radio communication is possible using a PC and TNC since it's "digital" (dit and dah). However, Morse code operation via computer is often a pain in the nether regions unless the other operator is also sending computer-generated Morse. If the code is being sent by a hand key, the resulting characters often have variations in spacing and time that differ from operator to operator. The TNC depends upon regularity in the Morse code characters for proper decoding, and the irregularities in hand-sent Morse makes it difficult or impossible for it to be decoded. As a result, hand-sent Morse often "prints" with large chunks missing in the middle of text or even as total garbage. On the transmit end, Morse code produced by a TNC has a lot of advantages over hand-sent code. The formation of characters and spacing between them is perfect, and it's possible to send code at precisely the speed desired. Thus, quite a few hams who operate CW with a TNC use the TNC for transmitting but copy the other station "by ear" if it's not also using a TNC to transmit CW!

Radioteletype (RTTY)—The length of a Morse code character varies from a single dit (the letter E) to lengthy combinations of dits and dahs, such as *dididahdahdidit* for the question mark (?). The dits and dahs of Morse code can be thought of as being equivalent to the bits making up a digital signal. RTTY uses a code, known as the *Baudot* code, in which all characters are of a uniform five-bit length. Unlike Morse code, the bits in Baudot are not determined by the length of the signal in time (like dits and dahs are determined) but instead by whether there is a signal present or not (i.e., "on" or "off" or 0 and 1). Baudot was originally developed for transmission of printed text over telephone lines, and operated like the old rotary dial telephones; the electrical pulses making up each five-bit character could be generated by a mechanical keyboard and used to drive mechanical printing systems. When RTTY first became used by hams in the early 1950s, stations were equipped with surplus mechanical keyboards and printers obtained from such companies as Western Union. With the introduction of the first personal computers in the mid-1970s, RTTY hams quickly embraced them as a way of getting rid of the noisy, bulky, and failure-prone mechanical units.

RTTY has some major limitations. Since five bits can only be arranged in 32 different ways, and there are already 26 characters in the alphabet, one five-bit character is defined as a "shift" character to enable transmission of numbers and some punctuation marks. Even with this, RTTY is restricted to all capital letters. RTTY is comparatively slow, with most ham RTTY communications carried on at a leisurely 45 baud (100 baud is the maximum baud rate allowed). And RTTY does not allow for error detection or correction nor for addressing of a message to a specific station. However, RTTY remains very popular on the HF bands. Since it's the oldest teleprinting mode, more stations are equipped for RTTY than other "print" modes. Operating procedures and conventions are well understood. If signals are strong enough so that error detection and correction is not a consideration, and you're just interested in having a one-on-one "conversation" with another station via keyboard, then RTTY is a fun and effective mode.

As we saw earlier, RTTY is sent by frequency-shift keying. The transmitter stays on continuously while the carrier is switched between the mark and space frequencies to form the various characters. On the VHF and UHF bands, an audio signal can switch between two fixed tone frequencies to accomplish the same thing, and this is known as audio frequency-shift keying (AFSK). At one time, RTTY using AFSK was popular on the VHF/UHF ham bands, but now has been totally eclipsed by packet radio.

American Standard Code for Information Interchange (ASCII)— ASCII is similar to Baudot, except each character consists of seven bits instead of five. These extra two bits permit 128 different characters, including lower case letters, extra punctuation, and special symbols. Otherwise, ASCII operation is similar to RTTY. And that's the problem—it's rare that hams need the extended character set offered by ASCII, so as a result ASCII operation is not too common.

Amateur Teleprinting Over Radio (AMTOR)—AMTOR is an extension of the Baudot code that incorporates powerful new error detection and correction features. Like RTTY and ASCII, AMTOR signals are sent using FSK or AFSK. Each character consists of seven bits, and each character must contain exactly four marks and three spaces. There are two types of AMTOR. *Mode A* can be used only between two stations. When two ham stations are in contact using AMTOR Mode A, the sending station is known as the *master* and the receiving station is the *slave*. These roles flip-flop as the two stations take turns sending and receiving, and the times at which they send and receive are synchronized. When the master station first contacts the slave, the slave responds with an acknowledgment signal back to the master indicating it is ready to receive. The master station will then transmit for a few milliseconds and pause for a few milliseconds. As the master transmits, the slave checks to see if each character has exactly four marks and three spaces. If the slave discovers a character in error, it will transmit an *automatic repeat request* (called an "ARQ") back to the master station during its silent period. Upon receiving the ARQ, the master station will retransmit the last data sent and will continue to do so as long as the slave sends an ARQ signal. The data received incorrectly will not be displayed until it is received correctly. When two stations are in contact using Mode A, they make a distinctive "chirping" sound like two birds singing back and forth to each other.

Mode A only works between two stations that are "locked together" in the master/slave relationship. Another method, *forward error correction* (FEC) or *Mode B*, is used for contacts between multiple stations or for information "broadcast" to any station listening. Instead of master and slave stations, each character is sent twice. If the first transmission of the character is received correctly, the second transmission of the character is ignored. If the second transmission of the character is also received incorrectly, a blank or error character is printed but no ARQ is sent to the transmitting station. Since the chances of noise or interference occurring during both transmissions of a character are low, Mode B offers a reasonable degree of pro-tection against transmission errors.

Neither Mode A nor Mode B is 100% error-proof, how-ever. Both systems only check whether a character has the necessary four marks and three spaces. It is possible for a char-acter to be received incorrectly but with four marks and three spaces. The result could be an incorrect character displayed on the receiving end, such as a "Q" printed when the sending station actually sent a "P."

AMTOR's use on the ham bands was slowed because it was introduced only shortly before workable implementations of packet radio became available to hams. Because of its greater error detection and versatility, more hams were attracted to packet radio. However, AMTOR has become recognized as a good approach for high reliability text transmission of the HF ham bands, where signal fading can cause problems for packet radio, and its use is now growing rapidly.

What's the "Packet" in Packet Radio Anyway?

The term *packet* refers to a way of organizing information. In other digital modes, each character transmitted is separate from the others and distinct in itself. In packet radio, data is grouped together into collections known as *frames*. A packet consists of one or more frames. In addition to the data we want to transmit, packets contain "flags" to indicate where each packet starts and stops, addressing information, and error-checking data. Figure 4-3 shows the organization of a typical packet and its individual frames.

Start Flag 8 bits	Address Field 560 bits Maximum	Control Field 8 bits	Protocol Identification Field 8 bits	Information Field 256 bytes Maximum	CRC Field 16 bits	Stop Flag 8 bits

FIGURE 4-3: Packets sound like random noise, but they're actually carefully organized into distinct patterns according to the protocol used.

The *start flag* of each packet alerts the receiving station that a frame of data is to follow and synchronizes the receiving station with the transmitting station, similar to the master/slave relationship found with AMTOR Mode A. The *address field* contains addressing information for both the sending and transmitting station. This addressing information includes the call signs of the sending and receiving stations and the calls of any stations that "digipeat" the packet to its final destination. When two stations are in contact by packet radio, they will ignore all packets not specifically addressed to them. The *control field* contains information about the type of packet being sent, including its number. Numbering of each packet helps with error detection and correction and helps the receiving station to assemble data transmitted with multiple packets into a correct sequence. The *protocol identification (PID) field* tells which data communications protocol is being used to transmit the packet. A *protocol* is a standard used to define the arrangement of data within a packet and its meaning according to its position within the packet. The most common protocol for packet communications is the AX.25 protocol, which is a variation of the X.25 protocol used for commercial data transmission over telephone lines. The *information field* within a packet can contain up to 2048 bits (256 bytes) of data. The CRC *field* is where each packet is checked for accuracy. CRC stands for "cyclical redundancy check," a mathematical technique which can generate a unique number when applied to the data bits in a packet. A CRC calculation is performed on the packet before it is transmitted and the result is added to the packet when it is transmitted. At the receiving end, another CRC calculation is performed and the result is compared to the number transmitted with the packet. If the numbers don't match, an error is assumed and the receiving station will request for the packet to be retransmitted. Finally, each packet is terminated with a *stop flag* that tells the receiving station that the end of the packet has arrived.

Whew!

(Don't worry. . . . you won't be asked all of this on the Technician class written exam. But these terms pop up constantly in articles, manuals, and books on packet radio, so I decided to introduce them here. After a while, these terms and the concepts they represent won't be so mysterious.)

Packets like the one shown in Figure 4-3 are sent as one continuous stream of data, and sound like a "brapp!" when heard as audio output from your receiver. The purpose of the AX.25 protocol is to divide the packet into the different fields and flags. With all stations using the same AX.25 standard, data can be sent as a continuous stream and the "division" into different fields will be done according to the AX.25 standard at the receiving end.

Packets aren't all alike, however. The two major types are *information* and *supervisory* packets. An information packet has an information field and is used to send information between two packet stations. A supervisory packet has the same fields as an information packet, except that it has no information field. Supervisory packets are used to control communications between two stations, such as sending a request for retransmission of a packet back to the sending station or to initially synchronize the transmitting and receiving stations.

One important point to keep in mind is that the information field of a packet is not restricted to character data. Anything that can be stored in binary form—and that includes audio and visual information—can be placed in the information field. Of course, the 256 byte capacity of a single information field is limited, but it's possible to break large amounts of data down into a series of individually transmitted packets. The packet numbering system lets the receiving station arrange the received data in the right order.

How Packet Radio Works

When two ham stations are communicating via packet, they are said to be *connected*. This is a carry-over term from communications over telephone lines using modems, but it's an appropriate description. When two packet stations are connected, they interact much like they would if there was a physical wire connection between them. Both the transmitting and receiving station use the same frequency in packet communications.

When you send a packet to a station you're connected to, the TNC at the other station looks for the start flag that tells it a packet is about to begin. The start flag also synchronizes the receiving station with the transmitting station. This synchronization process is known as *clock recovery*. After synchronization, the receiving station will perform an independent CRC calculation on the received packet and compare the resulting value to the CRC calculation performed by the transmitting station and included with the packet. If the CRC values match, the packet is accepted as valid and the data it contains is accepted for further processing by the TNC. The TNC will also signal the transmitting station that the packet was received correctly by sending an ACK (acknowledgment) signal back to the transmitting station. A separate ACK signal is sent for each packet correctly received. The synchronization between the transmitting and receiving stations means that ACK signals are sent at a certain time by the receiving station, and that the transmitting station will listen for an ACK signal at that same time. If the ACK signal for a packet is not received by the transmitting station, that packet is resent by the transmitting station. Once a packet has been received correctly and an ACK signal sent to the transmitting station, the receiving station will then assemble the packets in their numerical order, remove all but the data field from each packet, and display the resulting data on the PC monitor or otherwise process the data (such as saving it to a disk drive) as desired by the receiving station operator.

Most packet radio operators today enter data at their PC keyboard. This means data is never entered in a smooth, consistent manner. Operators pause different lengths of time between keystrokes, or may wait several seconds to think of something to input. To work around these problems, the transmitting station TNC and its controlling software will store all characters entered at the keyboard in a special part of the PC's memory called a *buffer*. When the buffer fills with enough characters to equal the size of the allowed packet data field, the buffer is "emptied," framed with the start and stop flags and other fields, numbered and CRC calculations done, and the resulting packet is transmitted. It's also possible to manually empty the buffer and transmit a packet by pressing the ENTER or RETURN key on the keyboard of the PC used in the packet system.

During RTTY, ASCII, and AMTOR, the carrier from the transmitting station remains on the air even when no information is being sent (as between characters being entered at the keyboard or while the operator pauses between words). In packet, the transmitting station remains off the air except when it is actually sending a packet. This means that a packet station is only on the air a fraction of the time used by other digital modes, and that several stations can share the same frequency as long as two or more stations don't try to use the frequency simultaneously. If two or more stations try to transmit a packet at the same time, the result is called a *collision*, resulting in none of the transmitted packets being received correctly. To determine when packet stations should transmit if a frequency is being shared with other stations, a method known as *carrier sense multiple access* (CSMA) has been adopted by hams. Under this system, each packet station monitors the frequency in use to see if a carrier is on the channel (they "sense" whether a carrier is present). If a carrier is there, no other station will transmit as long as the carrier is on the frequency. If a station wants to transmit a packet, it will wait until the frequency is clear. However, it is possible for two or more stations to be waiting for the channel to "clear" so they can transmit. When the channel is clear, the stations waiting to transmit their packets will begin to do so at the same time, resulting in collisions and no ACK signals sent by the receiving stations. When this happens under CSMA, a random time delay generator in each transmitting station will activate and cause their next attempt at retransmission to occur at different times. As a result, one of the stations will retransmit their packet first and "seize" the frequency. Other stations will sense the presence of the carrier, and wait their turn until the frequency is open.

Packets are "addressed" to specific stations. An *address* consists of the call signs of both the sending and receiving stations. The TNC control software handles the address creation when two packet stations are in direct contact with each other. If your packet is to be relayed ("digipeated") by other packet stations before reaching its destination, then the address will include the call signs of stations relaying the packet. When two stations are connected to each other, they will ignore all packets not addressed to them. If there is a collision between packets, the interfering packets are not recognized by the receiving station and are not displayed on the PC monitor used with the TNC. If you're connected to another station and several other stations are trying to use the same frequency, you won't see the packets they send. Instead, you'll notice delays in your packet communications. There might be intervals of a few seconds between your input of data at the keyboard and its transmission, for example, or an equivalent delay in getting a response from the station you are connected to.

If you're not connected to a specific packet station, you can view all packets being sent on a frequency by using the *monitor* function of your TNC. This lets you receive and display all packets being transmitted on a frequency even if they are not addressed to you. This points up a crucial difference between packet radio and telephone line modem communications: packet radio is not in any way "private." Anyone with a TNC and appropriate software can monitor your communications. When you're connected to another station, it often seems as if you and the other station are the only ones on the frequency, and it's easy to lapse into thinking your communications are private. They're not! Monitoring a frequency before operating on it is a good idea, as it lets you ascertain the level of activity and some of the stations active on the channel.

There are times when you may need to simultaneously communicate with more than one packet station, with all stations being able to copy each other's packets. Unfortunately, the robust error detection and correction features of packet are only available when two stations are connected to each other. If three or more stations are in communication with each other by packet, the packets do not have any error correcting features and stations view all packets using the monitor function of their TNCs.

When you finish communicating with a specific station via packet, you *disconnect* from that station. This is done by sending a supervisory packet containing a disconnect command to the station you are connected to. Once the disconnect command has been sent, you're free to connect with other packet stations.

How Fast Is Packet?

If you're used to communicating over telephone lines using a PC and modem, you're probably curious how fast a baud rate you can use on packet. The rate is less via radio than by telephone because of bandwidth and propagation considerations. As we mentioned earlier, the bandwidth of a signal increases as the amount of information it contains increases. Thus, the faster the baud rate, the greater the bandwidth of a packet signal. Radio communications are inherently less reliable than those over a telephone line, and their reliability decreases as the data rate increases. If reliability decreases to the point where frequent retransmissions of packets are required, the effective baud rate is lowered and the advantages of a higher data rate are negated.

On the HF ham bands below 30 MHz, signal fading and interference is common. At high baud rates, quite a few packets can be disrupted and retransmission required. As a result, HF packet is restricted to 300 baud. If you're familiar with modem baud rates used for telephone line communications, this might seem slow. Comparatively speaking, it is; however, it is a real improvement over the 110 baud rates typically used for HF RTTY and ASCII communications! On the VHF and UHF ham bands, FM is used to transmit packet and it has all the advantages of FM voice, including the capture effect and better signal-to-noise ratios. Higher baud rates are used, with 1200 and 2400 bauds being common. Some experimentally inclined hams working at the cutting edge of technology are even trying packet at 19,200 baud! As such speeds are refined and come into widespread use, transmission of "almost real-time" visual and other non-text data will become practical.

Basic Communication Via Packet

As you might suspect, operating via packet is a lot different from the voice and non-voice modes we've looked at so far. For one thing, a lot of packet communications is done computer-to-computer without any human operators!

If you operate packet, you'll become very familiar with this prompt on your PC monitor:

```
cmd:
```

This is the *command* prompt used in almost all packet communications software. It is from this prompt that you will enter all commands to control your packet communications.

Let's suppose you know that I'm operating my packet station (AA6FW) on a certain frequency and want to contact me. You would enter something like this on your PC:

```
cmd: c AA6FW
```

The c AA6FW command tells your computer that you want to connect your station to mine. Your computer will then send a packet addressed to my station to establish a connection. If my station successfully receives your packet and is available for communication, it will send a return packet to your station and this will appear on your PC monitor:

```
*** CONNECTED to AA6FW
```

From this point, our stations are connected and we can communicate. You'll see your own input and the data received from me displayed on your PC monitor. As you finish entering your input at your PC keyboard, press ENTER or RETURN to transmit it to me. On my end, I'll do the same thing. We need some way to let each other know when we've finished a portion of a message and are waiting for a reply, so the usual practice is to end each completed portion with GA (for "go ahead") or K (borrowed from Morse code operation), as in:

```
I'll be there tomorrow. Will you? GA
```

The GA lets the receiving station know that it's their turn to transmit.

While you can use almost any terminal communications program for packet, it really helps if you have a "split screen" program that displays received text and the text you send in separate screen areas. The reason for this is that it is often possible for a packet to arrive while you're still filling the buffer for your next packet. If it does, and you're not using a split screen terminal program, then the incoming packet can appear in the middle of your input text! While you don't have to worry about transmitting this text back to the originating station as part of your input, the effect can be very disconcerting and confusing. A split screen terminal program helps keep things straight. Most software intended specifically for packet use is of the split screen variety.

How do you end a contact? On most PCs (especially MS-DOS systems), you can return to cmd: by typing the letter C while holding the CONTROL key down. When cmd: appears, just type d (for disconnect) and your communication with the other station is terminated.

However, you might receive a message like this if you try to connect to my station:

```
Sorry, will be available tonight after 8 p.m.
de AA6FW
```

One of the great things about packet is its ability to provide unattended communications. In this case, I've instructed my packet station to acknowledge all attempts to connect to it with a message that I'll be there in person later that night.

How do you call CQ on packet? Well, strictly speaking, you can't since all packets have to be addressed to a specific station. However, most packet software will let you address a packet to a station with the call letters CQ. Other stations on the frequency monitoring all packets being sent can read your packet and connect to your station if they want to talk. A "CQ" on packet often goes something like this:

```
Hi! Anyone around for a chat? This is AA6FW
```

And this is enough to get a contact started if anyone is monitoring and wants to "talk" with you.

There's another type of "CQ" procedure popular in the early days of packet that you should avoid today except in unusual circumstances. This is *beacon* operation, in which you can instruct your packet station to automatically send a packet addressed to CQ at specified intervals. When packet radio first began and stations were few, this was a good way to stir up activity and see who (if anyone) was actually using the frequency. At today's levels of packet activity in most parts of the country, beacons just needlessly clog the frequency. There may be some good reasons to operate your station as a beacon, such as if you're in an isolated area of low population or are operating packet on a lightly used frequency. On a frequency like 145.01 MHz in most of the country, a beacon is totally unjustified and will only earn you the disdain of other operators.

Packet lets you screen who can and can't connect to your station. Suppose the operator of KR2H is a real jerk and you don't want to find yourself connected to him. Most packet software will let you instruct your system to ignore connection requests from stations you specify! In this case, you could configure your system so that KR2H would be unable to connect to your station regardless of how often he tries. KR2H would still be able to monitor any packets you might send to other stations, however, so he would know you're active and what your packets contain.

You can communicate with packet stations beyond your normal communications range by using a digipeater. Any TNC will let you configure your station as a digipeater, and many packet operators whose stations have wide coverage normally leave their stations in a digipeater mode when they're not operating the station themselves. Figure 4-4 shows how a digipeater works. The digipeater is located so that it can connect to two other stations that cannot directly connect with each other. Let's suppose the digipeater has the call sign KA5M, and the two stations unable to directly connect with each other are AA6FW and KR2H. (KR2H is still a real jerk, but so is AA6FW. They have to connect with each other; everyone else has them screened out.) To communicate with KR2H using KA5M as a digipeater, AA6FW would enter the following at his PC:

```
cmd: c KR2H v KA5M
```

which addresses a connection request to KR2H via (v) KA5M. If KA5M is currently in service as a digipeater, it will forward the connection request to KR2H and in turn will relay any acknowledgment of the request back to AA6FW. Once the connection is established between KR2H and AA6FW, KA5M will be "transparent" to both stations; KA5M will be relaying packets for both stations but its presence in the communications path will not be noticed by either station. The only clue that their signals are being digipeated to each other will be a slowing down in the speed of communications. It might take a couple of seconds longer for each station to reply to or acknowledge something sent to it.

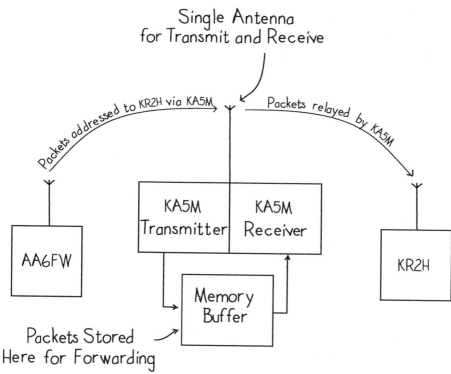

Single Antenna
for Transmit and Receive

Packets addressed to KR2H via KA5M

Packets relayed by KA5M

AA6FW

KA5M
Transmitter

KA5M
Receiver

Memory
Buffer

KR2H

Packets Stored
Here for Forwarding

FIGURE 4-4: Compare this digipeater to the FM repeater shown in Figure 3-11. The big difference is that a digipeater uses the same frequency for receiving and transmitting, while an FM repeater uses separate input and output frequencies.

A digipeater receives and transmits on the same frequency used by the stations relaying packets through it. When a station is operated as a digipeater, a *memory buffer* is set up in the available random access memory of the digipeater's PC. As a packet is received, its contents are sent to the memory buffer. When the buffer is full, the packet is then retransmitted to the station it is addressed to. This process is called *store and forward* operation, and is why communications through a digipeater are always slower than a direct connection between two packet stations.

Some digipeaters are operated strictly as digipeaters, much like voice repeaters; these are known as "dedicated" digipeaters." There are many other packet stations whose operators put them in a digipeater mode when they're not being used for ordinary packet communications.

Bulletin Boards and Mailboxes

If you're familiar with the concept of a telephone line bulletin board system (BBS), you already know how a packet BBS looks and works. A BBS is, in effect, another computer system open to all users. You can send messages to, and receive them from, other users of the BBS. These messages can be to specific stations or to anyone who "checks into" the BBS. This is like the e-mail familiar to users of telephone line computer communications systems. Other users can reply to your messages. Many bulletin boards have text files and software of interest to hams that you can *download* to your system. And, if you have created such material yourself, you can *upload* it to a BBS and share it with other hams.

An increasing number of bulletin boards are also incorporating store and forward facilities. These are linked with other packet bulletin boards, including some that operate on HF and even satellites. Even if you operate exclusively on VHF or UHF, it's becoming more common to be able to send and receive packet messages from other hams around the world by this method. The only catch is that you must know which BBS is the "home" BBS used by another ham to send a message this way. For example, AA6FW might have a BBS operated by KR2H as his home BBS. The address for packets sent to AA6FW through that BBS would be

AA6FW @ KR2H

which simply means "AA6FW in care of (or at) KR2H."

A *mailbox* can be thought of as your own personal BBS. This lets other stations leave messages for you—or receive them from you—even if you're not present at your station. A mailbox works like this. When you can't be at home, you set your packet station to operate in the mailbox mode and leave your station on. If another station tries to connect to yours, your packet mailbox will acknowledge the connection request and the other station can send packets to your mailbox, where they will be stored (either in buffer memory or to your disk drive) so you can read them when you return home. Some advanced TNCs go this one better and let you load messages addressed to various stations in your mailbox. If one of these stations connects to your mailbox, the appropriate messages will be forwarded to them! Some packet BBS stations will even forward messages they receive addressed to you to your packet mailbox.

Networking

The pathways by which multiple stations can communicate with each other constitute a *network* in packet radio. There are two types of networks of interest to us. A *local area network* (LAN) describes how packet stations within a narrow geographic area (such as a town or county) can communicate with each other. Most LANs make use of one or more digipeaters. Figure 4-5 shows how a typical LAN might work. Some stations can communicate directly with each other, while other stations must go through a digipeater or two. Each of the stations in the LAN is known as a *node* of the LAN. Regardless of the pathway, all of the stations in Figure 4-5 are able to maintain regular communications with each other. By the way, a station can be part of more than one LAN.

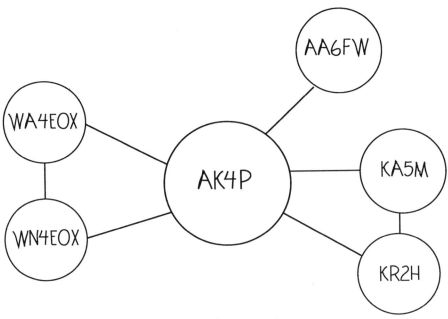

FIGURE 4-5: KA5M and KR2H can directly communicate with each other, as can WN4EOX and WA4EOX. Otherwise, all stations in this LAN have to communicate through AK4P. Gee, what a power trip for AK4P!

A *wide area network* (WAN) describes how LANs are interconnected to permit packet stations to communicate over a much larger geographic region, such as from the San Francisco area to southern California or Dallas to Houston. A LAN may be part of more than one WAN.

Not every packet passed in a LAN needs to be passed to a WAN. If you have a message to another station beyond your LAN, you access the WAN through a *gateway*. A gateway is a node of a LAN where messages can be sent to or received from the WAN. To keep interference down, most gateways use a different frequency to communicate with the WAN. For example, packet communications within a LAN may be conducted on 145.01 MHz. If a message intended for a WAN is by a LAN gateway on 145.01 MHz, the gateway will transmit the message to the WAN on another frequency such as 145.03 MHz.

WANs can be thought of as the packet radio equivalents of the airline travel system. Gateways are like the major "hub" airports, like those in Los Angeles and New York, that have numerous direct flights to each other. Airports in smaller cities such as Fresno or Santa Barbara have no direct flights to New York and are like nodes in a LAN. However, it is possible for passengers in such cities to fly to the Los Angeles "gateway" and thus travel to New York. In this example, the Santa Barbara to Los Angeles route is a LAN and the Los Angeles to New York route is a WAN.

Messages might be relayed by several different nodes within a WAN, and it often takes several minutes for a message to travel from the originating station to the destination station. This makes it difficult to carry on real-time conversations over WANs. Most messages forwarded by WANs involve nodes that are a combination of a BBS and gateway. A combination of HF and VHF frequencies can be used in WAN communications. Suppose you're in Los Angeles and want to send a message to your friend in Boston. You can address a message to your friend by indicating the BBS in Boston that your friend normally uses. You can send the message on 145.01 MHz to the BBS/gateway node of your LAN. When the message reaches the gateway, the gateway could send the message to the BBS in Boston on a HF packet frequency, such as 14.109 MHz. When the BBS in Boston receives the message on 14.109 MHz, it could store it for your friend or, if he or she has a mailbox TNC, even automatically send the message to your friend using a Boston LAN. In addition to HF frequencies, messages can be forwarded by satellites or chains of WANs operated strictly at VHF.

Packet Radio and the Future

Packet radio is still in its infancy! Prior to 1983, packet radio was the domain of hard core experimenters, and packet radio did not really get going until the introduction of the first affordable TNCs in 1985. The nationwide WAN system is still growing and forming, and what will emerge in just a few years will make today's networks look primitive by comparison. A lot of hams feel that packet radio is where the future of ham radio lies, and if you're a computer hobbyist or experimenter you can help write history.

Strange Packet Lingo

Even oldtimers in ham radio are baffled by some of the terms used by "packeteers." Here are some of the terms you need to know to be cool when it comes to packet radio.

address: the information that tells which other packet station a packet is intended for and the relaying path, if any, that it should be sent by.

attended operation: when a human operator is present and controlling a packet station.

baud: the speed at which data is transmitted measured in bits per second. 300 baud is the same thing as 300 bits per second.

beacon: packets addressed to any receiving station that are automatically transmitted at specified intervals by a TNC. Widely used in the early days of packet to alert other users of a frequency of stations available for communication, it's now a great way to irritate other packet users.

collision: when two or more stations attempt to send a packet at the same time on the same channel. This is known as "interference" to non-packeteers.

connected: when two stations are in contact by sending packets addressed to each other. The stations acknowledge correct receipt of packets from each other and do not recognize packets from other stations.

digipeater: a packet radio station that receives and retransmits packets intended for another station.

gateway: a node that's part of two or more networks and can send and receive messages between those networks.

mailbox: a TNC that can automatically receive and store messages received from a packet station or BBS.

monitoring: when a receiving station is displaying packets that may not be addressed to it and is not acknowledging correct receipt of those packets. This is a great way to "see" the activity in your area without having to transmit.

network: a method of interconnecting packet radio stations so that packets can be exchanged over long distances.

node: an individual station in a packet network.

PACSAT: a ham radio satellite having store and forward capabilities.

protocol: a standard set of rules and definitions describing how communications between packet stations will be conducted.

real-time: when communications seem to be happening without any perceptible delay, much like a face-to-face conversation.

store and forward: a packet system where messages can be received, stored, and then retransmitted at a future time.

unattended operation: when a packet station is performing certain operations, such as retransmitting packets addressed to another station, without a human operator present.

Propagation and Simple Antennas

ROPAGATION IS A TERM used to describe the process of how signals travel from one station to another station. Perhaps the most important item in determining how well this process works is a station's antenna. Together, propagation and antennas determine the distance over which you can communicate with your ham station. Let's see how each work separately and together to determine the distance over which you can normally expect to communicate.

How a Radio Signal Travels

A radio signal can take a lot of different paths from a transmitter to your receiver. Let's look at the three most important ones.

One case is where both the transmitting and receiving stations can "see" each other. This doesn't mean stations have to be in actual visual contact with each other (for example, you could be using a walkie-talkie inside your house), but only that a straight line can be drawn between the stations—much like shining a flashlight from one station to the other. This is known as *line of sight* propagation, and is most common on the VHF and UHF bands. Repeater communications are an example of line of sight communications; repeaters are on a tall building or mountain and can "see" a wide area (and, in turn, other stations can see the buildings or mountains where the repeaters are). A radio signal propagated by line of sight is called a *direct wave*. The usual range of a direct wave is the optical horizon as seen from the highest station plus about 20%. In addition, direct waves can reflect off buildings, mountains, and similar solid objects. Some hams even communicate by bouncing direct wave signals off the moon! (If you're interested in trying something like that, you'll need to use the maximum permitted transmitter power and a elaborate, highly directional antenna system.)

While limited in range, line of sight propagation by direct wave is usually highly reliable. The effectiveness of direct wave communication increases with higher frequencies, and it's the main propagation method on the ham bands above 50 MHz. Certain atmospheric conditions (which we'll discuss soon) can cause normal communications range at VHF and UHF to be extended, but the distance covered by the direct wave remains constant as long as the locations of two stations remain the same. The strength of signals propagated by direct wave follows the inverse square law, meaning that the strength drops to one-fourth of its original strength as the distance between stations doubles.

Other radio signals travel along the surface of the Earth from the transmitting station to the receiving station. This is known as *ground wave* (or *surface wave*) propagation. One key advantage of ground wave over direct wave propagation is that the ground wave can travel over hills and mountains that can block direct wave signals. Ground wave propagation can also extend well beyond the optical horizon and is more effective at lower frequencies; the ground wave will travel much further at 3.5 MHz than at 144 MHz. In fact, daytime signal propagation on the AM broadcast band (540 to 1600 kHz) is almost exclusively by ground wave. (However, the ground wave is present both day and night.) The exact distance covered by the ground wave will depend upon the surface it travels over, with salt water and moist soil giving greater communications range than rocky ground. However, distances of about 100 miles or so can be reliably covered by the ground wave emitted from a transmitter operating at 4 MHz or below. But on the 10-meter band (28 to 29.7 MHz) the ground wave typically covers less than 20 miles. On the higher UHF bands, the ground wave only travels a few feet before fading out! Figure 5-1 will help you visualize how the direct and ground wave work at different frequencies.

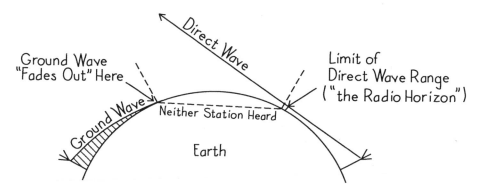

FIGURE 5-1: The ground wave, as its name implies, travels along the surface of the Earth until it fades out due to the inverse square law. The direct wave can be thought of as "line of sight" propagation—if you can "see" the other station (even if it's a couple of hundred miles away), you can communicate with it.

The third type of radio signal is one that travels from the transmitter up to outer space. This is known as the *sky wave*. You might think the sky wave is wasted transmitter energy, since there aren't many hams in outer space. But actually the sky wave is a very important propagation method. Above the stratosphere is a region of the Earth's atmosphere known as the *ionosphere*. In the ionosphere are various layers electrified (*ionized*) by energy from the Sun. These layers can bend (or refract) sky wave signals back to Earth, where they can be received thousands of miles away from the transmitter site. This phenomenon affects many ham bands, so let's look at it in more detail.

About the Ionosphere

The ionosphere varies in its height above Earth, as heating by the Sun causes it to expand and contract. However, it begins at an altitude of 25 to 55 miles (40 miles typically) and continues to approximately 250 to 400 miles above the Earth. The ionosphere is divided into the following layers:

D layer: This is the lowest layer, existing at altitudes of approximately 25 to 55 miles. This layer is usually present only during the daytime. The lack of energy from the Sun causes this layer to fade away shortly after sunset.

E layer: The layer is found from 55 to 90 miles above the Earth. Like the D layer, this layer is mainly present only during the daytime, although it takes longer to fade away after sunset. However, this layer often produces conditions that allow communications on 50 and 144 MHz over distances of hundreds or thousands of miles. (We'll discuss this phenomenon later.)

F layer: This is the most important layer for sky wave communications. The F layer begins at about 90 miles and continues upward to over 400 miles. During daytime (and particularly in summer), the F layer can split into two separate layers known as the F1 and F2 layers. At night, only a single F layer is present. Sky wave signals refracted by the F layer (or layers!) can be received thousands of miles away. The F layer is the main propagation method for DX signals on the HF bands, and also provides exciting DX on the 50 MHz band during years of high solar activity.

Figure 5-2 shows how the ionosphere above your location varies at your local noon and midnight during a typical summer day.

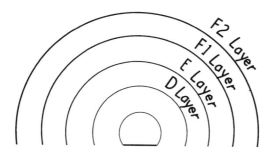

A: Ionospheric Layers at Local Noon During Mid-Summer

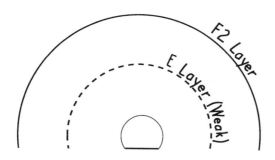

B: Ionospheric Layers at Local Midnight During Mid-Summer

FIGURE 5-2: The ionosphere is like the ocean; it changes throughout the day.

The ionosphere becomes electrified mainly due to the effects of ultraviolet radiation from the Sun. When ultraviolet radiation from the Sun reaches the atoms of the gases making up the ionosphere, it "ionizes" them—in other words, it causes the atoms to gain an extra electron. The ionized atoms in the ionosphere act like a sort of "electric mirror" capable of bending sky wave signals back toward the Earth. The ionosphere's ability (or inability) to refract signals, as well as the frequency of the signals that can be refracted, depend on the level and type of solar activity. When signals are refracted by the ionosphere, they are sometimes described as having "skipped" off the ionosphere. *Skip* is a general term used to describe any sky wave propagation by the ionosphere.

The ionosphere doesn't treat all sky wave signals the same. Depending on the frequency of a signal and the time of day, the ionosphere may refract it back to earth, absorb it, or let it pass through into outer space. For example, let's suppose it's your local noon on July 14. A sky wave signal at 1800 kHz (the lower edge of the 160-meter band) will be totally absorbed by the ionosphere and not refracted back to Earth. A signal on 21 MHz (the 15 meter band) will be refracted, usually over distances of hundreds or thousands of miles. However, a signal on 432 MHz (the 70-centimeter band) will penetrate through all layers of the ionosphere into space. At your local midnight on December 25, however, things will be different. The 1800 kHz signal will be refracted back to Earth then, while the 21 MHz signal will pass through the ionosphere into space like the 432 MHz signal does. 50 MHz (6 meters) is the highest frequency ham band which can be refracted by the F layer.

The different layers of the ionosphere treat radio signals differently. The D layer is the weakest layer of the ionosphere; indeed, it's so weak that most radio signals just zip through it on their way to the E and F layers. However, the D layer usually absorbs some energy from signals passing through it and weakens them. This absorption is especially pronounced on lower frequency signals below 10 MHz or so, and increases as the signal frequency drops. The strength of the D layer over your location is greatest at your local noon and really drops as sunset approaches; it vanishes soon after sunset. During some short midwinter days, the D layer may not form at all.

Like the D layer, the E layer is primarily a "daytime only" layer—reaching greatest strength at local noon—and absorbs low frequency signals. However, it is stronger than the D layer, absorbing more of a low frequency signal, and can remain present during the early evening. However, very intense patches of ionization can form in the E layer at certain times of the year. This is called *sporadic-E* and the patches can refract 28, 50, and 144 MHz signals back to Earth at distances of up to 1200 miles. Sometimes a signal will be refracted off multiple sporadic-E "clouds" and distances of over 2500 miles can be covered. In North America, sporadic-E is most common from late May to early August, with another brief peak around the winter solstice. Sporadic-E usually occurs in the mid-morning and late afternoon to early evening, although it can happen at any time of day or year. A sporadic-E opening can be a wild experience! It's not unusual for 6 meters to be dead, only to spring to life a few minutes later with stations hundreds or thousands of miles away. Since the sporadic-E clouds are in motion, the "direction" of an opening can change abruptly. You might be able to work stations in western New York with ease, and then suddenly the opening will favor stations in eastern Virginia. Stations can fade out just as unexpectedly as they can fade in.

Sporadic-E is a very common event on 10 and 6 meters each year, and there are a handful of openings on 2 meters as well. It has been reported on extremely rare occasions on 222 MHz, but is generally believed to be impossible on any higher frequencies. Sporadic-E can last for hours on 10 and 2 meters, with openings on 2 meters being much shorter (often only a few minutes). If you get a code-free Technician license, one of the reasons you'll look forward to summer will be the return of 6 meter sporadic-E DX!

Strange Propagation Lingo. . . .

As with everything else, hams have their own special language to describe events involving propagating signals from one point to another. Here are some of the more common ones used on the air and in this book.

blue whizzer: a meteor that produces a signal that can be heard for an exceptionally long time, such as two consecutive minutes.

burst: a brief snippet of signal heard via meteor scatter, usually lasting only a few seconds at most.

closed/closing: the end of propagation on a frequency. "I was working him fine on 10, but the band just closed on us in the middle of the QSO." Or "I'll sign with you now, I think the band is closing on us." (This is a handy way to get out of a boring contact.)

dead: a frequency or band will not support desired propagation. "I just checked 15 meters before coming here; boy, is that band dead!"

E-skip: another term for sporadic-E propagation.

flutter: when a signal is rapidly varying in strength in a rhythmic pattern, usually because it is being propagated by a path over one of the polar regions.

folded: another way of describing what happened when a band closed.

LUF: the *lowest usable frequency,* and the opposite of MUF. Signals lower than the LUF will be absorbed by the ionosphere and can't be propagated by sky wave. LUF varies likes the MUF.

multihop: a radio signal that is refracted more than one time by the ionosphere before reaching the receiving site. This can often be recognized by a fading, "echoing" signal.

opening: when there is propagation on a certain frequency or band. "There was a great tropo opening to Indiana and Illinois last night on two!"

path: the route a radio signal takes from the transmitter to eventual receiver.

ping: another term for a burst.

short skip: ionospheric propagation over no more than a few hundred miles.

tropo: another term for tropospheric ducting.

The F layer is the one responsible for most DX on the HF bands. During periods of high solar activity, it can also refract 50 MHz signals. During daytime in much of the year, the F layer splits into separate F1 and F2 layers. Most signals that can penetrate the E layer also penetrate the F1 layer, so most daytime HF propagation is by the F2 layer. At night, the F1 layer weakens and merges with the F2 layer to form a single F layer. However, during many midwinter days there will be only a single F layer present due to the shorter days and reduced solar energy reaching the ionosphere.

How the Sun Affects the Ionosphere

The level of ionization in the ionosphere above your head varies during the day. It's usually lowest just before your local sunrise. At that time, the D and E layers have long faded away and only a single F layer is present. As the sun climbs in the sky, the ionization begins a steady increase and the D and E layers re-form. At your local solar noon (when the sun reaches its highest point in the sky that day, not when the clock says it's noon), the ionization of all layers is at its highest. As the sun moves westward, the level of ionization begins a slow decline. After sunset, the D layer quickly fades away and is followed soon afterwards by the E layer. The level of ionization in the F layer declines throughout the night, reaching a minimum just before sunrise the next morning. At sunrise, the process starts over again.

If you think the ionization would be greater in the summer than the winter, you're right. Not only are the days longer, but the sun's rays are more direct. (Since the seasons are reversed in the northern and southern hemispheres, the ionization effects are also reversed.) But the results might be different than you expect. During the winter, the ionosphere is heated less by the Sun. As a result, the F layer does not expand as much as it does in the summer. The result is a single F layer which is densely ionized despite the reduced amount of energy received from the Sun. At night in winter, the F layer rapidly loses ionization. During the summer, the ionosphere receives more energy from the Sun. But the increased heating spreads out the ionosphere over a larger area, splitting it into the F1 and F2 layers, but the resulting average ionization density is often less than during midwinter days! After sunset, though, the F layer retains more ionization in summer than in winter as the F1 and F2 layers contract into a single, densely ionized layer.

Perhaps the most important factor affecting ionization is the level of sunspot activity. The Sun emits more ultraviolet radiation when it has more sunspots, and as a result the ionosphere is more heavily ionized.

Sunspots appear on the Sun in a series of cycles. A *sunspot cycle* is defined as the period from a minimum number of sunspots to a peak number and back down to a minimum again. Cycles take several years to complete, with 11 years being an average cycle. However, sunspot cycles have been observed lasting as few as 5 years or over 17 years. Reliable records of sunspot activity and cycles have been kept since the mideighteenth century. At maximum, the sunspot count of some cycles has exceeded 150 (a sunspot cycle which peaked in March, 1958 had a maximum count of over 200) while the count during a cycle minimum can be as low as 10. The average maximum is usually around 110.

A high number of sunspots isn't good news for all HF bands, however. As the number rises, so does the absorption of lower frequency signals by the various layers of the ionosphere. Frequencies at or below 10 MHz (the 160, 80, 40, and 30 meter bands) actually experience better sky wave propagation during years of low sunspot numbers. These same bands are usually better for sky wave propagation in the winter than in summer, again due to reduced absorption of those signals by the different layers of the ionosphere.

The Maximum Usable Frequency (MUF)

The highest frequency the ionosphere can propagate by sky wave is known as the *maximum usable frequency* (MUF). This is also referred to as the *critical frequency*. This varies with the time of day, season of year, and the number of sunspots.

As a general rule, the daytime MUF will be higher in winter than in summer. At night however, the reverse is true: night MUFs are higher in summer than in winter. The ultimate determinant of the MUF is the sunspot cycle. During years of low sunspot numbers, the maximum daytime MUF in winter might not even reach 20 MHz. But during years of high sunspot activity, the daytime winter MUF can easily exceed 50 MHz, meaning the 6 meter band is open to most areas of the world for several hours at a time.

If a signal exceeds the MUF, it will pass through the F layer and be lost into space. The closer a signal's frequency is to the MUF, the better it is propagated. A very low power transmitter can communicate over surprising distances when it is operated near the MUF.

Skip Zones and Dead Spots

When you get on the air, you'll notice there are some puzzling gaps in what you can hear via sky wave propagation. For examples, signals over 1500 miles away might be roaring in at very strong S9+ levels, while signals from a few hundred miles away will be extremely weak or completely inaudible. Common sense would say that stations closer to you should be stronger rather than unreadable. What's going on?

Figure 5-3 shows what is happening. The sky wave signal travels upward to the ionosphere, where it is refracted back toward the Earth. But the sky wave signal can't be heard "under" the refraction point—it's literally "over the heads" of potential listeners until it skips back down. In between the transmission and reception points, there is a large dead spot beyond the ground and direct wave range of the transmitter. This spot where the signal cannot be heard despite being closer to the transmitter than the eventual reception point is known as the *skip zone*.

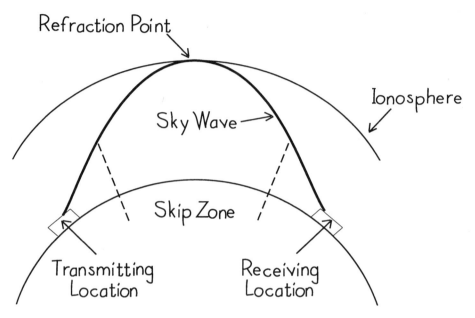

FIGURE 5-3: Our friend the ionosphere in action! This is how long distance radio propagation works via the E or F layers of the ionosphere. The "skip zone" is an area where the signal can't be heard (it's "skipping" overhead) despite being closer to the transmitter site than the eventual receiving location.

The skip zone between two stations isn't a constant. It varies with frequency, time of day, and season of the year. If a station you want to communicate with by sky wave is located in the skip zone, there is one possible method you can try. It's called *scatter*, and involves using the highest possible transmitter power to "scatter" some radio energy back into the skip zone from the overhead refraction point. Sometimes scatter works because some of the energy reaching the distant receiving site will be refracted back into the skip zone when it makes contact with the Earth at the receiving site. Scatter is basically little more than throwing a lot of sky wave energy upward and hoping that some will come down in the area you want it to. Scatter signals are usually weak in the skip zone even if the maximum legal transmitter power is used. However, it's often the only way to work stations in a skip zone on the HF bands.

Tropospheric Ducting Propagation

The 6 meter band (50–54 MHz) is the highest band where sky wave propagation via the F layer can take place. Propagation via sporadic-E is common on 6 meters and happens only occasionally on two meters. The other VHF and UHF ham bands are unaffected by the ionosphere. However, the VHF and UHF bands experience a special type of propagation unknown on frequencies below 50 MHz. This is known as *tropospheric ducting*, and can result in ham stations operating above 50 MHz being able to conduct communications over hundreds or even thousands of miles.

The troposphere is the part of the atmosphere that's closest to Earth, extending from the Earth's surface to approximately 6 miles. This is where weather happens; rain, wind, and storms are all products of the troposphere. Normally, the troposphere has no effect on radio signals of any frequency. But this can change when certain weather conditions are present. The temperature of the troposphere usually drops as altitude increases. But sometimes a layer of cool, dry air close to the Earth's surface is overridden by a layer of warmer, more moist air. This means that, at a certain elevation, the air temperature actually *increases* as the altitude increases. This situation is known as an *inversion* and is commonly found along weather fronts or large bodies of water. The warmer, more moist air acts as a "shield" to signals above 50 MHz, and bends them along the inversion until it ends. Signals can travel hundreds or even thousands of miles until the inversion ends. Figure 5-4 shows how tropospheric ducting works.

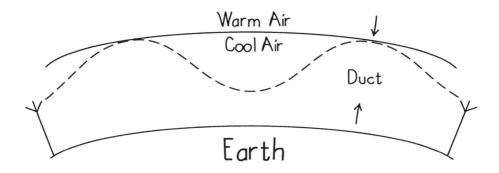

FIGURE 5-4: Tropospheric ducting "traps" VHF/UHF signals close to the Earth's surface and prevents them from escaping into space.

The effects of tropospheric ducting increase with frequency. It's more noticeable on 2 meters than 6 meters, and even more so on 70 centimeters than 2 meters. An inversion can last for days, and although signals are not as strong as during a sporadic-E opening they are usually more stable. Inversions are common during the fall months along slow moving fronts where cool, dry Pacific air meets warm, moist air from the Gulf of Mexico. Indicators of conditions favorable for formation of inversions are warm, hazy days followed by cool nights; inversions also trap pollutants close to the Earth, so smoggy or foggy conditions often indicate an inversion is present.

Tropospheric ducting over less than 300 miles is also fairly common during summer and fall around sunrise. This happens as the Sun first warms the upper part of the troposphere before the lower; such openings usually vanish within an hour of sunrise.

Ducts generally cannot form over mountainous areas, so hams in Rocky Mountain states experience less tropospheric DX than the rest of the country. Ducts across the Rockies (as from California to eastern Colorado) are extremely rare. However, the Appalachian Mountains are not high enough to seriously disrupt an inversion, so tropospheric ducting is common between any two points in the United States east of the Rockies. Hams along the Pacific coast enjoy frequent ducting openings up and down the coast; some hams in California have even been able to contact Hawaii on 2 meters and higher bands through inversions!

Other VHF/UHF Propagation Modes

There are two other propagation modes that work on frequencies above 50 MHz. Both are based on phenomena more typically associated with astronomy than radio.

One mode is *meteor scatter*. When small meteors enter the Earth's atmosphere, friction with the atmosphere causes the vast majority of them to burn up before reaching the ground. (If you've seen a "shooting star" at night, this is what you've seen.) As they travel through the atmosphere and vaporize, such meteors leave behind a trail of ionized gas and particles. These ionized trails can be thick enough to refract signals at VHF and UHF frequencies. Meteor scatter is most prevalent at 50 MHz, less so at 144 MHz, and is rare at higher frequencies. The bursts produced by meteor scatter can last from only a few seconds to over two minutes. While meteors are random phenomena, there are several times each year when a large number of meteors can be reliably expected, usually as a result of the Earth's orbit taking it near some cometary debris. These are known as *meteor showers*. Several VHF/UHF enthusiasts like to make schedules for contacts during meteor showers. Since the exact time when meteor scatter will happen is unpredictable even during a meteor shower, stations during such scheduled contacts transmit and listen at precisely scheduled times. For example, a station might transmit beginning on each minute for 15 seconds and again for 15 seconds beginning at 30 seconds after each minute; the rest of the time, that station would listen for the station it is scheduled to be in contact with. This can be a slow process, but with enough meteor bursts a complete contact can be made.

Another phenomenon that aids VHF propagation is the *aurora*. Solar flares can release large amounts of ultraviolet radiation and charged particles. These charged particles are attracted by the Earth's magnetic field and enter the ionosphere at the North and South Poles. These charged particles produce the brilliant auroral displays seen in northern regions of the northern hemisphere and also severely disrupt the ionosphere over the polar regions. (Often, HF signals that cross the polar regions will have a distinctive flutter on them as a result of the disturbed ionosphere.) The disrupted ionosphere may be sufficiently ionized to propagate VHF signals, however. The best band for auroral propagation seems to be 2 meters. If you can see a visible aurora on the northern horizon from your location, odds are that auroral propagation on VHF is possible. One interesting quirk about auroral communications involves the orientation of directional antennas. As we'll soon see, you would normally orient a directional antenna according to where a station you desired to contact was located. If you wanted to contact a station to your west, you would orient a directional antenna so that its maximum response and power output was toward the west. However, in auroral conditions one would orient a directional antenna to the *north*—toward the ionized region over the poles. Suppose a station in Oregon wants to contact a station in North Dakota via auroral propagation. Instead of orienting their antennas toward each other, both stations would instead aim their antennas north toward the auroral region.

What "Makes" a Good Contact?

A "quickie" propagation mode like meteor scatter raises an interesting question: what makes a good contact—that is, a contact where we're sure we heard the other station and we're sure they heard us? This is a question that often arises in contests and DX operating as well.

There's no FCC rule covering this, but most hams agree that an acknowledgment of the other station's call sign and a signal report is enough to constitute a valid contact. Something like the following is all you need:

"AA6FW, QRZ?"

"AA6FW, this is KR2H. You're five by nine."

"Roger, KR2H. You're also five by nine. QSL?"

"QSL and thanks. KR2H out."

"QSL. AA6FW, QRZ?"

Short and sweet—and utterly cryptic to non-hams. But also undeniably a valid contact. Both stations clearly heard each other, exchanged signal reports, and acknowledged correct receipt of the information.

Which HF Bands Are Good for What and When

The HF bands are 160 through 10 meters. If you get a Novice or Technician-Plus license, you'll be able to operate on parts of 80, 40, 15, and 10 meters. If you get a plain vanilla Technician license—or don't get a license at all (but why would you do something that stupid???)—you're still welcome to listen to these bands with a shortwave receiver. Let's look at the typical propagational characteristics of the HF bands.

160 meters: The band begins just 200 kHz above the upper end of the AM broadcast band, and it is very similar in propagation. Daytime propagation is via ground wave and is generally limited to 100 miles. At night, sky wave propagation takes place over typical distances of a few hundred miles in summer to thousands of miles during the winter. This band offers better sky wave propagation during years of low sunspot activity.

80 meters: This band has propagation very similar to 160 meters, although sky wave propagation is usually better. Quite a few hams have worked over 100 countries here, and DX contacts of a few thousand miles are common in winter. This band is also better for sky wave propagation during years of low sunspot activity.

40 meters: This is a transition band with many interesting characteristics. During summer days, it offers reliable communications over 300 miles or so. On winter days, this range can be extended to over 500 miles. Night communications at distances of well over 1000 miles are common, with better conditions during winter frequently allowing intercontinental communications. Unlike 160 and 80 meters, most communications here are via sky wave. The main problem with this band is the interference from powerful international shortwave broadcast stations located outside of North America. If you operate on frequencies without such QRM, however, this is a terrific band for reliable night communications throughout the United States and Canada. Forty meters is better for communications during years of low sunspot activity, although the improvement is not nearly as dramatic as on 160 and 80 meters.

30 meters: This band is much like 40 meters, although it is limited to CW and RTTY communications and does not suffer interference from international broadcasting stations. This band offers better day and night range than 40 meters, and is an especially good choice for reliable daytime range out to about 1000 miles or so. This band is usually better at times other than the peak of a sunspot cycle; the increased ionospheric absorption of low frequency signals can reduce the strength of signals on this band.

20 meters: This is the main band for DXers. During years of low sunspot numbers, there are frequent daytime openings of several thousands of miles. During years of high sunspot activity, the band is often open around the clock to some distant part of the world! During the summer, there are often good evening openings to the west (where the sun is still shining) or to areas to the east where sunrise is taking place. Communications on 20 meters is normally via sky wave, although direct wave communications within a few dozen miles is possible. This holds true for 17, 15, 12, and 10 meters as well.

17 meters: This band is very similar to 20 meters, although daytime openings are fewer during years of low sunspot numbers.

15 meters: This band is also much like 20 meters, although it is much more influenced by the sunspot cycle. At the minimum of a sunspot cycle, this band may not open even during the daytime. At the peak, 15 meters is often better than 20 meters for DX communications.

12 meters: This band is heavily influenced by the sunspot cycle. During years of low sunspot numbers, it is good only for local direct wave communications. During years of high sunspot activity, this band is open during the daytime for DX communications over several thousands of miles. It is also often open during sunspot cycle peaks for DX with stations located to the west during the evening hours. When the MUF is just above this band, low powered stations using simple antennas are capable of worldwide communications with ease!

10 meters: This is like 12 meters, only more so. During years of low sunspot activity, this band is dead for any sort of propagation by the F layer. Communications can be carried on by direct wave over distances of about 25 miles (much like the range of CB radio) and there are often sporadic-E openings of a few hundred miles similar to those found on 50 MHz. When sunspot numbers are high, daytime and early evening DX on 10 meters is spectacular! A very simple, low powered station can easily work over 100 countries during the months centered around a sunspot cycle peak. Stations and countries that are difficult or impossible to work on 20 meters during a sunspot cycle minimum can be easily worked on 10 meters during a sunspot cycle peak.

Which VHF/UHF Bands Are Good for What and When

If you get a codeless Technician or higher class license, you'll have full access to every one of these bands. If you get a Novice license, you'll be able to use parts of 1.25 meters and 23 centimeters. The VHF and UHF bands have seen substantial growth in the amount of usage and types of activities carried on them in recent years, and the codeless Technician license promises to keep that trend going.

6 meters: This is really a schizophrenic band. During years of low sunspot activity, it is a strictly local band like 10 meters. However, there are many summertime sporadic-E openings even at the bottom of a sunspot cycle. At the peak of a sunspot cycle, it's a different story; some hams have worked over 100 countries on 6 meters! Tropospheric ducting is also found here, although it is not as noticeable or effective as on higher frequencies. This is the highest frequency ham band at which sky wave propagation via the F layer is possible.

2 meters: This is the most popular ham band in the United States (and the world, for that matter). Repeaters help make this the optimum band for local phone communications, and most packet activity is found here. Sporadic-E is sometimes found here. A clue that sporadic-E will be possible on 2 meters can be found by monitoring a 6 meters sporadic-E opening;

when an intense opening on that band suddenly "shortens" so that stations within 300 or 400 miles are being heard, that's a good indication that the E layer ionization is becoming intense enough to support an opening on 2 meters. Tropospheric ducting is very common here, and DX via meteor scatter and aurora is often conducted using CW just above 144 MHz. If you could operate on only one ham band, 2 meters would be the one to go with!

1.25 meters: This band is similar to 1.25 meters for local communications, although it is much more lightly populated. If you like local communications on 2 meters but don't like crowds, this band is a good alternative. Sporadic-E, meteor burst, and auroral propagation are very rare. However, tropo is even more common than on 2 meters.

70 centimeters: This is the first band that is generally considered UHF instead of VHF, and is the home of repeaters, packet networks, television, ham satellites, a host of linking and control signals, and such esoteric communications modes as bouncing signals off the moon. Normal, "non-repeatered" local communications are over shorter distances than on the VHF bands, and signals here more readily are reflected by such objects as steel frame buildings and mountains. Some hams take advantage of this property by orienting their directional antenna toward the building or mountain and allow it to reflect signals in the desired direction! This increased "reflectivity" of signals means this band (and those of higher frequency) are useful when you are using a walkie-talkie inside a building; the signals will bounce around inside the building with little loss of strength until they find a way out of the building through a window or air shaft. Tropospheric ducting is common here.

33 centimeters: This is a relatively new UHF ham band and is still lightly populated; it attracts hard-core experimenters who like to build their own equipment and antennas. At these and higher frequencies, signals become more "light-like" and can be more easily reflected or blocked by metal structures. The reflective qualities present on 70 centimeters are enhanced, but there are also more "dead spots" where a building, mountain, or other object will reflect a signal back to its source, producing a "radio shadow," and prevent communications between two points. Tropo is also found here.

23 centimeters: Signals at these frequencies become even more like light and highly reflective. This is another popular band for satellite and television communications, and its width (60 MHz!) makes it popular for experimenting with new modes. At these frequencies, possible injury due to body tissue heating is a real problem. Prolonged exposure at high power levels to signals at these frequencies can have effects similar to those produced by a microwave oven! The biggest danger is when using a walkie-talkie operating on this band. If such a walkie-talkie is held near the eyes and used extensively, it's possible to cause injuries to the eyes. Fortunately, a few simple precautions (which we'll discuss later) can prevent such injuries. Tropo is found here, but local communications can be spotty due to reflection from different objects and radio shadows from structures.

There are other UHF ham bands, but they are largely experimental at this point. Unless you build your own equipment and antennas, 23 centimeters is about as "high as you can go."

Antenna Basics

An antenna is a metallic object used to send and receive radio signals. When electrical energy from a transmitter is applied to an antenna, it is converted into electromagnetic energy which is radiated as radio waves. When receiving, the process is reversed. When radio waves (electromagnetic energy) strike the antenna, it sets up faint electrical energy in the antenna that the receiver can then amplify and convert into sound or some other form. There are two basic types of antennas. The *omnidirectional* or *non-directional* antenna radiates or receives radio signals equally well in all directions. By contrast, a *directional* antenna will favor certain directions over others and is usually designed so that it can be rotated or the direction of maximum radiation changed. Figure 5-5 shows the *radiation patterns* of typical non-directional and directional antennas as seen from above the antennas. Note that the directional antenna pattern is actually "multidirectional." The area of greatest radiation strength is called the *major lobe* while the areas of reduced radiation are called the *minor* or *secondary lobes*. By the way, these patterns also apply to receiving as well as transmitting. A directional antenna helps you reject interfering stations outside the antenna pattern. This directivity in reception, coupled with the concentration of transmitter power in a desired direction, means that a directional antenna vastly increases the communications effectiveness of any ham station. It also means that if you're trying to increase the range over which you can communicate it's usually a smarter idea to use a directional antenna rather than just increase transmitter power. In this chapter, we'll concentrate on the principles of non-directional antennas. The theory behind directional antennas is a little more complex, and we'll put off discussing them until Chapter 7.

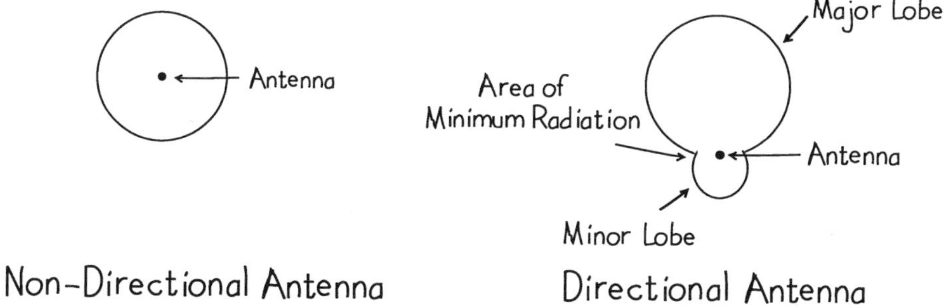

Non-Directional Antenna   **Directional Antenna**

FIGURE 5-5: Directional antennas let you put more transmitter power where you want it put!

We've noted in previous chapters that the size of antenna needed for a particular frequency varies *inversely* with the frequency—the higher the frequency, the smaller the size of antenna required. The basic type of antenna is known as a *half-wave antenna*, which means its physical size is approximately one half of a wavelength long at the operating frequency (that is, a half-wave antenna for the 20 meter band would be about 10 meters long). This is because an antenna of the proper length for a given frequency lets the electrical energy from the transmitter travel from one end of the antenna and back within one cycle. Back in Chapter 3, we discussed how wavelength is the distance between the same points on two consecutive radio waves and how wavelength varies with frequency. Thus, an antenna that is one half wavelength long is the right length to let the radio energy travel back and forth its entire length during a single cycle of a radio wave. When an antenna is at the right length for a certain operating frequency, it is said to be *resonant* at that frequency.

A half-wave antenna is "fed" energy from the transmitter in its middle. As we mentioned back in Chapter 2, energy is transferred to the antenna using either a parallel conductor or coaxial cable feedline. When the feedline is connected, it divides the half-wave antenna into two quarter wavelength segments, as shown in Figure 5-6. If parallel conductor feedline is used, one of the conductors is attached to one quarter-wavelength segment and the other conductor is connected to the remaining quarter-wavelength segment. If (as is usually the case) coaxial cable is used, the center conductor is connected to one quarter-wavelength segment and the metal shield "braid" is attached to the other quarter-wavelength section. The antenna shown in Figure 5-6 is the most basic and most widely used antenna on the HF bands. It's called a *dipole*.

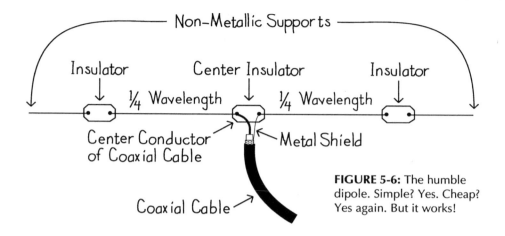

FIGURE 5-6: The humble dipole. Simple? Yes. Cheap? Yes again. But it works!

There's a simple formula for determining the correct length of a dipole (half-wave) antenna for a given frequency, as follows:

length in feet = 468/frequency in MHz

I can almost promise you that you'll have at least one question on your written exam requiring you to use this formula! For example, a typical question might be to calculate the length in feet of a dipole antenna intended for use on 7150 kHz. To solve this problem, you would first convert the frequency from kHz (7150) to MHz (7.15). Then you just plug 7.15 into the formula and divide:

468/7.15 = 65.45 feet

If a dipole is "cut" for a specific frequency (sometimes called the center frequency), it might seem like a very limited antenna. However, a dipole will operate properly and with high efficiency over a wider frequency range than the single frequency it is physically cut to. For the HF ham bands, this is usually about 150 kHz above and below the center frequency. In this example, a dipole cut for 7150 kHz will provide good results throughout the 7000 to 7300 kHz 40-meter band. If you try to operate a dipole more than approximately 200 kHz away from its center frequency, however, the performance will drop off and you'll have problems getting your transmitter to "load" properly into it. If your transmitter can't be loaded correctly into your antenna, a lot of your transmitter power won't be radiated and some is reflected back from the antenna to the transmitter, possibly damaging your transmitter. We'll talk about antenna loading in more detail in the next chapter.

A dipole is not an omnidirectional antenna. Instead, it has a "figure-8" radiation pattern as shown in Figure 5-7. The dipole in Figure 5-7 will concentrate most of the transmitter power to the east and west and also receive best from those directions. For dipoles, this effect increases with frequency. As a general rule of thumb, it makes little difference which direction you have a dipole oriented on the 160, 80, and 40 meter bands. While there will be noticeable directional effects on those bands, they will not be significant under most conditions. But on higher frequency bands they will be more important and must be kept in mind when locating and installing a dipole.

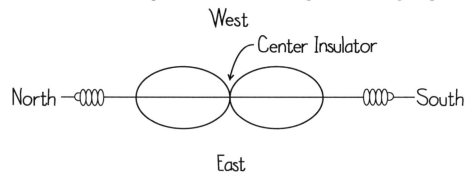

FIGURE 5-7: A dipole radiates in a "figure-8" pattern. This pattern gets "tighter" and more pronounced as the operating frequency is raised. On 80 and 40 meters, it doesn't really matter which direction you place a dipole. But on higher bands the directional effects are more significant.

A dipole is a *balanced* antenna. This means that both "sides" (i.e., the two quarter-wavelength sections of the dipole) are symmetrical around the "feed point" of the antenna (as shown in Figure 5-6) and neither side is connected to *ground*. The term "ground" pops up frequently in radio and electronics, and we'll be using it a lot in the rest of this book. Ground originally referred literally to a connection to the Earth itself, usually through a metal stake driven into the ground. This practice originated in the early days of electricity and radio to provide a way to discharge dangerous voltages, such as lightning strikes. "Ground" still refers to such direct connections to the Earth, but also can mean a connection having similar properties as one to the Earth (such as to the metal hull of a ship at sea), any point of zero voltage in a circuit, or a common connection point for any part in an electronics circuit. In this chapter, we'll use ground to mean a connection to the Earth or its equivalent.

Now take a look at Figure 5-8. There's a quarter-wavelength vertical section perpendicular to the ground above the feed point, but where's the other half of the antenna? The other "side" of the antenna is the ground itself! This is why a vertical antenna is known as an *unbalanced* antenna. The "ground" side can be the Earth itself, a series of metal wires called *radials*, the metal body of a car or ship, or even sea water. The more closely the ground approximates the qualities of the Earth itself, the more effective the vertical antenna will be. To find the correct length of a vertical antenna, perform the same calculation as for a dipole antenna and divide the result by 2.

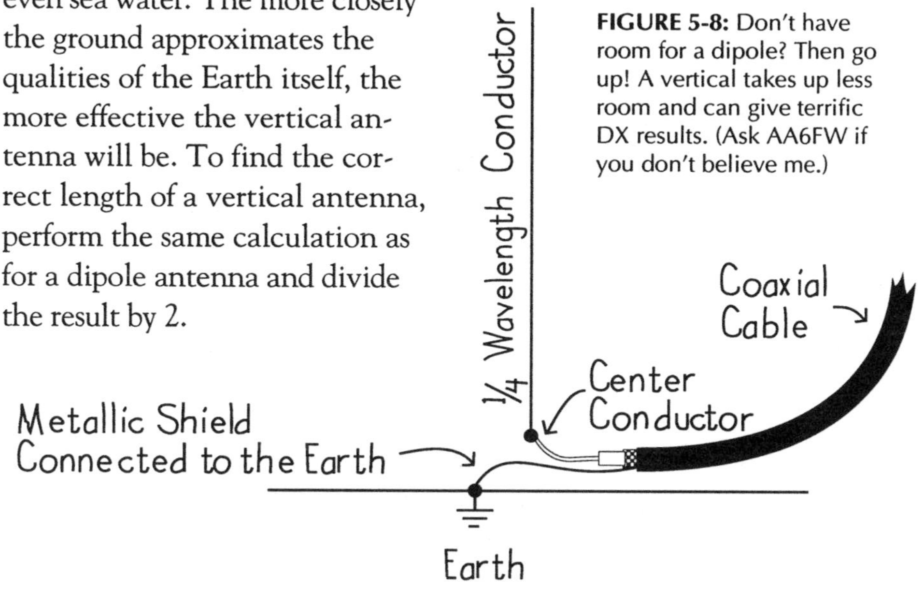

FIGURE 5-8: Don't have room for a dipole? Then go up! A vertical takes up less room and can give terrific DX results. (Ask AA6FW if you don't believe me.)

¼ Wavelength Conductor

Coaxial Cable

Center Conductor

Metallic Shield Connected to the Earth

Earth

Polarization

So far in this book, we've talked about radio waves without going into what they really are. Radio waves are waves of *electromagnetic* energy. The term "electromagnetic" means radio waves consist of interlinked electric and magnetic *fields*; these different fields are at right angles to each other. Like light, radio waves are also *polarized* and the position of the electric field determines the polarization of a radio wave. If a radio signal is said to be horizontally polarized, this means the electric field of the wave is traveling parallel to the surface of the Earth.

The polarization of a radio wave when it's transmitted is determined by the physical position of the antenna. A horizontally mounted antenna, such as the dipole in Figure 5-6, will radiate horizontally polarized waves. The vertical antenna in Figure 5-8 will radiate—you guessed it!—vertically polarized waves, which means the electric field component of the wave will be perpendicular to the Earth. We could change the dipole in Figure 5-6 to a vertically polarized antenna merely by installing it in a vertical rather than horizontal position.

The polarization of a radio wave can be altered after it's transmitted by refraction in the ionosphere or by reflection off objects at VHF/UHF frequencies. Most sky wave signals have a polarization that changes constantly.

On the HF bands below 30 MHz, polarization is not a major factor. There are a lot of arguments over whether a horizontally or vertically polarized antenna is best below 30 MHz, but no clear winner has been conclusively established. However, many hams prefer horizontal polarization because most electrical noise (such as from power lines and electrical devices) is vertically polarized. This is true even when the noise sources, such as power lines, are physically horizontal. On the VHF and UHF bands, antenna polarization is crucial. At those frequencies, horizontally polarized antenna will have difficulty in "seeing" a vertically polarized wave and vice-versa. This effect increases with frequency; it's more pronounced on 432 MHz than 50 MHz, for example. By common agreement, most ham activity on VHF and UHF takes place with vertically polarized antennas. This is mainly done because of the large number of FM rigs installed in cars and walkie-talkies on the VHF/UHF bands. For such stations, a vertically polarized antenna is the only practical alternative despite being more susceptible to noise (would you want to cruise down the highway with an antenna sticking out from both sides of your car???).

Loading Coils, or How to Get an Antenna to Think It's Something It's Not

You may know some hams who have a HF transceiver installed in their car. The antennas mounted on the rear bumpers of their cars are only a few feet high. Since we've seen that the length of an antenna varies with its frequency, how can an antenna only a few feet long be resonant on a band like 80 meters?

Fortunately for us, it's possible to electrically "fool" an antenna into thinking it is longer than is actually the case. This is done by adding a *loading coil* to the antenna structure. As we'll see in the next chapter, a loading coil is an example of an "inductance." For now, just remember that the loading coil causes the effective *electrical* length of an antenna to be longer than its actual physical length. So why don't hams use loaded antennas all the time, such as for their home stations? One reason is that the bandwidth over which a "loaded" antenna is resonant is much more narrow than a full-size antenna—sometimes it's as little as 20 kHz on a band like 80 or 40 meters. A more important reason is that we might be able to trick the antenna and our transmitter into thinking that the antenna is the right physical length, but Mother Nature isn't fooled. Some of the transmitter power is lost as heat and "leakage" from the loading coil. A loaded antenna will radiate some signal, but not as much as an equivalent physically full-sized antenna for the same frequency range. Nonetheless, a loaded antenna can give good results and permit operation in circumstances where a full-sized antenna for a certain band is an impossibility.

Stretching Antennas

It's common, particularly on the higher frequency bands, for vertical antennas to be longer than a quarter wavelength. A very popular "stretched" antenna is 5/8-wavelength long (or, since we're talking about a vertical antenna, *high*) at its operating frequency. The reason is because such antennas have more *gain* than quarter-wavelength antennas for the same frequency. "Gain" means that the signal from a 5/8-wavelength antenna seems stronger than the same signal radiated by a quarter-wave-length antenna. How is this possible?

Figure 5-9 shows the radiation patterns of a quarter wavelength and 5/8-wavelength antenna. The quarter-wave-length antenna has a hemispherical pattern; as much energy is being radiated straight up as it is to

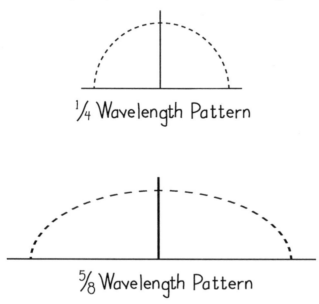

¼ Wavelength Pattern

⅝ Wavelength Pattern

FIGURE 5-9: Lengthening a vertical antenna "flattens out" the radiation pattern.

the sides. Compare that to the 5/8-wavelength pattern. The "overhead" radiation of the antenna is greatly reduced, and more of the energy is radiated toward the horizon. This is especially desirable at VHF and UHF, where sky wave propagation is minimal and radiation upward is just wasted energy. The effect of concentrating more energy toward the horizon is the same as increasing the transmitter power applied to a quarter-wavelength antenna, and so a 5/8-wavelength antenna is said to have more gain than a quarter-wavelength antenna.

We'll talk more about antennas, particularly directional types, in Chapter 7. However, we've gone about as far as we can go without getting into some nitty-gritty details about electricity and electronics. We'll take a look at those in the next chapter.

CHAPTER SIX

Electrons, Capacitors, and Other Stuff That Makes Your Radio Work

Considering how pervasive electricity and electronics are in our daily lives, it's remarkable how little most people really understand about them. Electronics has produced so many near-miraculous things that the subject might seem as if it must be impossibly difficult to understand. If you're new to electronics, you might be dreading having to learn about what goes on inside a radio and all the different parts that make it work.

Relax. Electronics can look really scary when you first get into the subject, but that's because so much of it involves stuff that's unfamiliar to you. Nothing that you'll need to know for your Novice or Technician written exam is very difficult—it's all well within the ability of an average junior high school student to grasp. (And don't forget that some elementary school students have also passed written exams for ham licenses!) The biggest barrier to overcome is the notion that electronics is "supposed" to be difficult. The material in this chapter only looks hard—it's really easy! Approach it for what it is, take it a bite at a time, and you'll do fine.

What Is Electricity?

Everybody knows "about" electricity, but just *what* is it? To answer that, we have to go down into the atom. As you learned in school, an atom consists of a nucleus, which is mainly *protons*, surrounded by *electrons* which revolve around the nucleus at different levels above the nucleus. (This is really too simplified, but it's just the protons and electrons we're interested in for our purposes.) It's the electrons whizzing around in the *outermost* level that makes electricity possible.

Protons have a *positive* charge while electrons have a *negative* charge. If two particles with the same charge (like two electrons) encounter each other, they repel each other with a very strong force. But two particles with opposite charges—an electron and proton—attract each other with a very strong force. In an atom, these particles and their charges are delicately balanced. This balance is what holds an atom together.

However, sometimes an atom can lose an electron from its outermost level. When this happens, it seeks to add a "replacement" electron for the one it lost. Until it does, the atom has more positive charge (thanks to its protons) than it does negative charge (because it's missing an electron), and so the atom is said to be *positively charged*. The reverse can also happen; an atom can pick up an extra electron in its outermost level and have more negative charge than positive. Such an atom is said to be *negatively charged*.

Atoms hate to be left in either a positively or negatively charged state, and are always looking for some way to find a missing electron or get rid of an extra one. Take a look at Figure 6-1. At part A, we see a negatively charged atom near a positively charged atom. The extra electron will be attracted from the negatively charged atom and move to the positively charged atom, as shown in part B. When this happens, the charges in both atoms will again be equal, as shown in part C.

A: Negatively Charged Atom Near Positively Charged Atom

B: An Electron Is Attracted to the Positively Charged Atom

C: Charges In Both Atoms Are Now Equal

FIGURE 6-1:
Love among the atomic particles? Or just a momentary attraction? You be the judge.

Electricity is a "flow" of electrons caused by atoms seeking to equalize their positive and negative charges. The flow starts at a place with negative charge (excess electrons), and moves toward a point with negative charge (too few electrons). This flow of electrons is called electric *current*, and it's measured in *amperes*. Actually, the term "flow" is a little misleading in describing what's going on in an electric current. A single electron doesn't start at the negatively charged point and move all the way to the positively charged point. Instead, the effect is more like standing dominos on end in a straight line and then tipping over one at the end of the line; the other dominos fall as they get hit by the one next to them. An electric current flow is more like a "ripple" through the wire or other material the current is flowing through.

Another way to describe the state of charge at a point is as its *polarity*. This is a term borrowed from the familiar north and south "poles" of a bar magnet. A negative polarity means a point has excess electrons, and a positive polarity means it has a deficit of electrons.

Some materials have electrons in the outermost levels of their atoms which are easily displaced and can move to other atoms. An electric current can move easily in such materials, and these are known as *conductors*. Examples of conductors are copper, gold, and aluminum. In other materials, the electrons in the outermost level are tightly "bound" to the atom and do not move easily to other atoms. These materials are called *insulators*. Glass, rubber, plastic, porcelain, wood, and air are all good insulators. Other materials, such as carbon, fall in between these extremes and let current flow but with some opposition, known as *resistance*.

While it's common to speak of "generating electricity," we don't really produce new electrons when we do so. The excess electrons already exist. When we generate electricity, we just make those electrons move through a conductor. Often, the negatively and positively charged points will be separated by a considerable "distance" (on the atomic scale of things!) and we need some sort of force to get electrons moving from the negative point to the positive one. This force is called *electromotive force* (EMF) or *voltage*. It's measured—appropriately enough!—in *volts*.

A lot of people get current and voltage confused. The best analogy is to think of the water flowing through a garden hose. The number of gallons (or liters, for all you fans of the metric system) of water flowing through the hose at a given moment is like current. The pressure that moves the water through the hose is like voltage. The hollow interior of the hose is like a conductor, and the rubber or plastic exterior of the hose is like an insulator.

If we were to increase the water pressure within the hose enough, it would eventually burst. This is why so many people equate "high voltage" with danger from electricity, since a sufficiently high voltage can break down an insulator and force current through materials that would normally insulate. However, low voltage electricity of high amperage can be just as dangerous.

Units such as amperes and volts are often too big for many purposes. Sometimes we need to refer to currents and voltages in terms of *milliamperes* (mA) and *millivolts* (mV). The prefix "milli" means that the unit is divided by 1000. Thus, 1 ampere = 1000 mA and 1 volt = 1000 mV. Even smaller units are the *micro-ampere* (μA) and the *microvolt* (μV). These are equal to one-millionth of an ampere and volt, respectively. Just remember:

$$1 \text{ ampere} = 1000 \text{ mA} = 1,000,000 \text{ μA}$$
$$1 \text{ volt} = 1000 \text{ mV} = 1,000,000 \text{ μV}$$

Electric Fields and Magnetic Fields

As you probably remember from your school science classes, electricity and magnetism are linked. If a conductor has electric current flowing through it, that conductor is surrounded by a magnetic field. And if a conductor is moved through a magnetic field, an electric current flows through the conductor. This is because electricity and magnetism are actually two different manifestations of the same force, known as *electromagnetism*. In the last chapter, we mentioned how radio waves are actually waves of electromagnetic energy and that radio waves consist of alternating electric and magnetic fields. But what exactly are electric and magnetic fields?

A good working definition of a field is a region of space in which a certain force, such as magnetism or the attraction between opposing polarities, is felt. Another way to think of a field is as the "area of influence" of a force. A force can influence something within its field, much like a magnet attracts iron filings sprinkled near it. A force like magnetism is often described in terms of *lines of force*. This is just a handy way of describing the "shape" of a field and how strong the force is at different places in the field.

Electricity needs to travel through a conductor, but forces can travel through the air or even a vacuum (like radio waves do). For example, suppose you place two copper plates parallel to each other but not touching each other in any way. If you were to connect one plate to the positive terminal of a battery and the other plate to the negative terminal of the battery, an *electric field* will be set up between the two plates. The opposite charges will attract each other, but the excess electrons of the negatively charged plate will be unable to travel across to the positively charged plate. However, the force will create the electric field between the two plates. In a similar fashion, a magnetic field will exist between north and south magnetic poles that are near each other without touching.

The Inverse Square Law

The strength of electric and magnetic fields—as well as electromagnetic and gravity fields along with the strong and weak atomic forces—is determined by something called the *inverse square law*. This says that the strength of a field varies according to the mathematical square (that is, a number multiplied by itself) of the distance between the source of a field (like a ham transmitter) and the point where the strength of the field is determined (or received, like a distant station you're trying to work).

In practical terms, let's suppose that you double the distance between a radio transmitter and receiver. The inverse square law says we take the square of doubling (i.e., 2) the distance, meaning the field is only one-quarter as strong at the new distance as it was before. Now let's suppose that you move the receiver to a new position that's three times closer to the transmitter than before. The strength of the field at the new position is not three or six times stronger—instead, it's *nine* times stronger!

The inverse square law was first stated by Sir Isaac Newton in his *Principia Mathematica* back in 1687. Newton developed the law in connection with gravity, but the brilliance of his discovery is shown by the fact that the law applies to things Sir Isaac never envisioned, such as radio waves and the forces that bind together subatomic particles. Think any of today's "breakthroughs" will still be valid over 300 years from now, or applicable to phenomena we can't even conceive of today?

Direct Current (DC), Resistance, and Ohm's Law

When we set up a path for current to flow through, we have a *circuit*. Figure 6-2 shows a simple circuit which is nothing more than current flowing from the negative terminal of a battery to the positive one. Since this current flows only in this direction, it's called *direct current* (DC).

Notice that the circuit shown in Figure 6-2 has a funny "squiggly line" symbol in it. This is the schematic symbol for a *resistor*. A resistor is a circuit component that opposes the flow of electric current through it. The unit by which we measure resistance is the *ohm*, whose symbol is Ω. It's also common to measure resistances in *kilohms* (1000 Ω, usually abbreviated k) and *megohms* (1,000,000 Ω, usually abbreviated M).

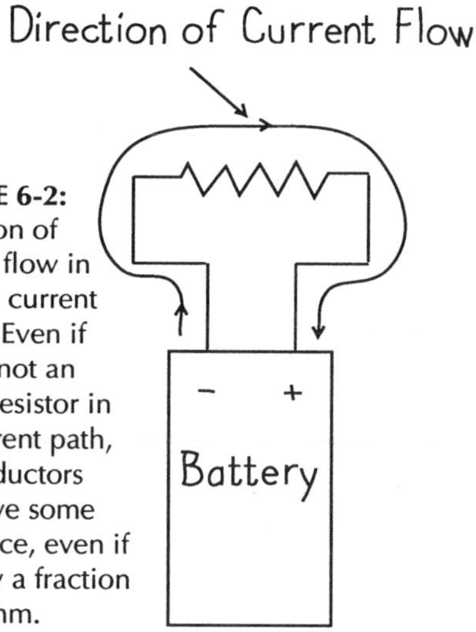

Direction of Current Flow

FIGURE 6-2: Direction of current flow in a direct current circuit. Even if there's not an actual resistor in the current path, all conductors will have some resistance, even if it's only a fraction of an ohm.

Resistance isn't a property related just to resistors. *All* conductors—even the very best ones, such as gold and copper—have at least some minor resistance to current flow. But resistors let us precisely specify the amount of resistance we want to use in a circuit. Resistors come in a variety of shapes and forms, and some (known as *potentiometers*, or "pots") let us vary the amount of resistance they have. (Volume controls are typical situations where we need variable resistors.) However, the most common type of resistors are "fixed" types, such as shown in Figure 6-3. These resistors are cylinder-shaped, have a pair of metal leads for attaching the resistor to other components, and are marked with three or four color bands. These bands are a *color code* that indicate the value of the resistor in ohms. The fourth color band (if any) indicates the *tolerance* of the resistor. The tolerance is how closely the actual value of the resistor is to the value indicated by the first three color bands. For example, a gold fourth band means the actual is within 5% (high or low) of the indicated value. If there is no fourth color band, then the actual value of the resistor is within +/- 20% of the indicated value.

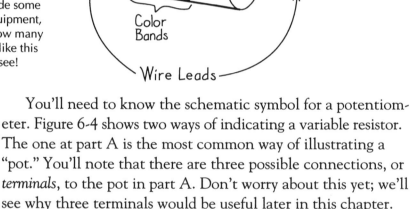

FIGURE 6-3: The next time you have a chance to look inside some radio equipment, notice how many resistors like this you can see!

You'll need to know the schematic symbol for a potentiometer. Figure 6-4 shows two ways of indicating a variable resistor. The one at part A is the most common way of illustrating a "pot." You'll note that there are three possible connections, or *terminals*, to the pot in part A. Don't worry about this yet; we'll see why three terminals would be useful later in this chapter. Part B shows another possible way to illustrate a variable resistor. This method is often used when we want to show a pot with only two terminals.

FIGURE 6-4: These two potentiometer symbols will soon become very familiar to you.

A: Three-Terminal Potentiometer

B: Two-Terminal Potentiometer

Resistors are made from such materials as a carbon composition, a carbon film, a metal film, and wire wound over a form ("wire-wound" resistors). Most resistors are made from carbon composition.

One important way resistors can vary is in their *power dissipation* rating. Resistors get warm as they oppose the flow of current due to internal "friction" between the atoms of the resistor and the flow of electrons moving through the resistor. In some devices, such as electric lights and heaters, we try to enhance and make use of this heating effect. However, in most cases the heat from a resistor is an unwanted (but unavoidable) by-product. The amount of heat dissipated by a resistor is measured in watts, and as a general rule the physical size of a resistor of a certain value increases as its power dissipation rating increases. For most electronics work, carbon composition resistors with power dissipation ratings of 1/8 or 1/4 watt are common, although carbon composition resistors of up to 2 watts are available. Wire-wound resistors can have power dissipation ratings of up to several hundred watts. For resistors of the same ohms value, you can freely substitute one of a higher power dissipation rating for one of lower dissipation with no problems. However, if you try to substitute a resistor with a lower power dissipation rating for one with a higher rating, the lower rated substitute will likely be burned up in short order!

Resistors can be connected together in two basic ways, *series* (one after another) and *parallel* (side by side). Part A of Figure 6-5 shows resistors connected in series and part B shows resistors connected in parallel. In the series connection, the individual resistors are connected one after the other so that all of the electric current will flow through each resistor. The total resistance of resistors connected in series is easy to calculate: just add up the values of the individual resistors. For example, the total resistance of three resistors connected in series could be:

$$17 \text{ k} (17{,}000 \, \Omega) + 520 \, \Omega + 2.2 \text{ M} (2{,}200{,}000 \, \Omega) = 2{,}217{,}520 \, \Omega$$

The parallel connection arrangement is much different from the series one. In part B of Figure 6-5, there are several different paths the current can take through the circuit and its resistors. We won't go into the theory of how it happens, but for now just remember this: in *a circuit where all the resistors are in parallel, the total resistance of the circuit is always less than the resistance of the lowest resistor value*. (Don't ask me why this is so, because I don't think I can explain it too well; just trust me that it's so!) To find the total resistance of a circuit with three or more resistors in parallel, such as shown in part B of Figure 6-5, add the *reciprocals* of each resistor together, and then divide the resulting sum into 1, according to this formula:

$$\frac{1}{\left(\dfrac{1}{R1}\right) + \left(\dfrac{1}{R2}\right) + \left(\dfrac{1}{R3}\right)} = \text{total resistance of resistors in parallel}$$

You'll have to get out your calculator if you have a problem like this on your written exam! But remember that the answer *must* be a smaller resistance value than that of the smallest value resistor. This can serve as a good check of the accuracy of your calculation.

Sometimes you'll have a situation like that in part C of Figure 6-5, where we have just two resistors connected in parallel. The formula for determining the resistance of just two resistors in parallel is:

$$\frac{R1 \times R2}{R1 + R2} = \text{total resistance of two resistors in parallel}$$

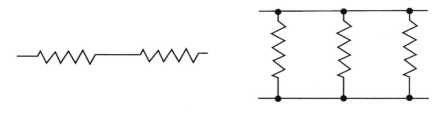

A: Resistors in Series   B: Resistors in Parallel

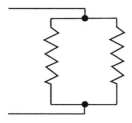

C: Two Resistors in Parallel

FIGURE 6-5: Here are how resistors can be connected in series and parallel. You'll probably see one of these three configurations on the written exam in connection (pardon the pun) with a question on using Ohm's law.

Resistance, current, and voltage are all related through *Ohm's law*. If we know the values of two of those factors, we can compute the remaining one using Ohm's law. This is a fundamental law of electricity, and perhaps the most important law as far as DC circuits are concerned. Ohm's law is expressed as three formulas, which are given below. You might as well memorize them, since there are questions about Ohm's law on both the Novice and Technician written exams and it also shows up on the other written ham exams. In these formulas, *E* represents the voltage in volts, *I* the current in amperes, and *R* is the resistance in ohms, as follows:

$$E = (I \times R)$$

$$I = (E / R)$$

$$R = (E / I)$$

Now let's look at some real-world examples of how these formulas are used in a circuit similar to that in Figure 6-2. These examples are taken from the Novice and Technician written exam questions in use when this book was written.

Finding a Resistance

Let's suppose we're using a 12 volt battery. Connected to it is a resistance that's drawing 0.25 ampere of current from the battery. What's the value of the resistance?

Our equation to solve this would be R = (12 / 0.25). Doing the division would give 48 Ω, which is the correct answer.

Finding the Current

Let's suppose the resistor in Figure 6-2 is 4.7 kilohms and that the battery voltage is 120 volts (yeah, it's a pretty damn strong battery!). How much current is flowing through the resistor?

There are a couple of places here where you can get tripped up if you're not careful. The first one is if you don't convert kilohms to ohms, so remember that 4.7 kilohms = 4700 Ω. Once this conversion has been made, the equation would then be I = (120 / 4700). Doing the division would give an amperage of 0.02553191489. This should be rounded up to 0.026 amperes. On the written exam, the correct multiple choice response for this could be either 0.026 amperes or 26 mA. Remember the milli/micro conversions!

Finding a Voltage

Let's suppose the resistor in Figure 6-2 is 50 Ω and that a 2 ampere current is flowing through it. What voltage is being supplied by the battery?

This is a really simple example since no unit conversion is involved. The equation to solve it is E = (2 x 50), giving a voltage of 100 volts across the resistor.

A very common application of this part of Ohm's law involves *voltage dividers*. While the Novice and Technician written exams in use while this book was written don't test you on this topic, it's such an important concept you should be familiar with it anyway. Voltage dividers make use of resistors to supply a desired voltage at a point in a circuit. Take a look at Figure 6-6. A circuit like this produces a fraction of the battery voltage at Vout. The formula for computing Vout is:

$$\text{Vout} = \text{Vin} \left(\frac{R2}{R1 + R2} \right)$$

Remember our discussion of potentiometers when we mentioned how many potentiometers are three-terminal devices? The circuit in Figure 6-6 and the schematic symbol part A of Figure 6-4 are very similar. This is no coincidence, since three terminal potentiometers are normally used in situations where we need to divide a voltage over a wide range of possible voltages.

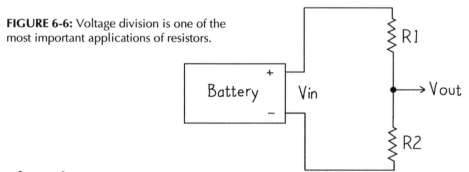

FIGURE 6-6: Voltage division is one of the most important applications of resistors.

Electric Power

Power is defined as the rate at which energy is being consumed in a circuit. Power is measured in *watts*, a term we've used earlier in this book without precisely defining it. Let's do so now.

Power is computed using a formula very similar to Ohm's law, in which power (P) takes the place of resistance (R) and voltage is represented by E and current by I:

$$P = E \times I$$

Just like Ohm's law, you can find the value of current or voltage if you know the value of the power and the other factor:

$$I = (P/E)$$

$$E = (P/I)$$

In earlier years, the maximum transmitter power permitted under FCC rules was determined by measuring the voltage and current in the last (or final) tube or transistor used to amplify the signal, and then computing the power. Since this was the power applied, or "input," to the last transmitter stage, this was known as *input* power. Nowadays things are a lot simpler. The maximum allowed transmitter power is specified as the power measured at the antenna terminal of the transmitter or power amplifier, or *output* power. To measure this power, just connect a device known as a *wattmeter* to the antenna terminal and read the output power directly.

Batteries

The schematic symbol in part A of Figure 6-7 is for a *single cell* battery. Single cell batteries are like the individual batteries that go in a flashlight or radio. The single cell batteries that go into a flashlight are also known as *dry cells*, and have a voltage of approximately 1.5 V. You'll often see the symbol shown in part B of Figure 6-7 used for batteries, however. This symbol represents a *multiple cell* battery. Multiple cell batteries can be two or more single cell batteries used together (such as the batteries powering a flashlight) or two or more single cells contained inside a single "housing" and having a common positive and negative terminal for all the cells. Note that in both diagrams the positive terminal is the leftmost "long line" and that a " + " is placed next to it to indicate the positive terminal. On the written exams, you might be asked to distinguish between the symbols for single and multiple cell batteries; such a question will sometimes contain a trick answer where a " – " is placed next to the leftmost long line, so be careful!

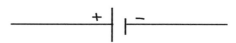

A: Single Cell Battery

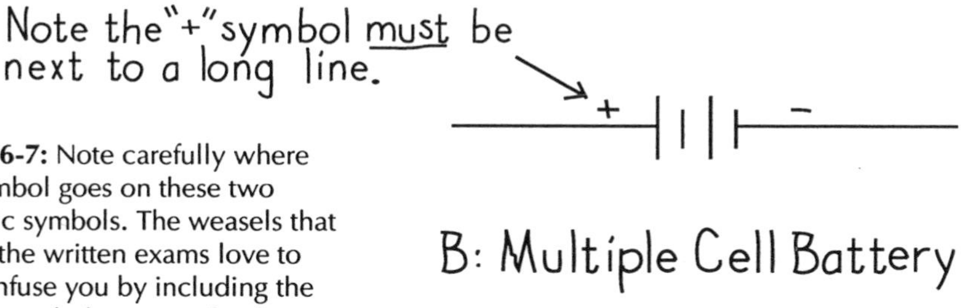

Note the "+" symbol <u>must</u> be next to a long line.

FIGURE 6-7: Note carefully where the + symbol goes on these two schematic symbols. The weasels that develop the written exams love to try to confuse you by including the identical symbol—except for the location of the + —twice among possible answers to a question!

B: Multiple Cell Battery

Open and Short Circuits

This is probably a good place to precisely define what we mean by an *open circuit* and a *short circuit*. An open circuit is one in which no electric current can flow. For example, we might send so much current through a resistor or other component that it is destroyed. The destroyed component would not be able to conduct any longer, the current wouldn't have a complete path to flow through, and the circuit would no longer work. Some components are designed to deliberately destruct when a certain current level is exceeded, and these are *fuses*. Fuses protect electronic circuits by creating an open circuit whenever certain dangerous levels of current occur. The schematic symbol for a fuse is shown in Figure 6-8.

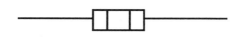

FIGURE 6-8: Schematic symbol for a fuse. Real exciting, huh? But fuses have saved a lot of ham radio gear from an early death.

The opposite of an open circuit is a short circuit. A short circuit is one in which current flows *too* well! In a short circuit, an unintended path of low resistance is created and large amounts of current flow through that path, creating a condition that's dangerous for circuit components and—more importantly!—humans. A short circuit can arise through the failure of a component or, more commonly, through careless wiring or assembly of a circuit. A short circuit can mean that an exposed part of an electronic device, such as the metal cabinet containing it, can have lethal current flowing on it.

If a component fails inside a circuit "catastrophically"—that is, the component is totally destroyed—you'll often hear it "pop" when it fails, smell something burning, or see thin wisps of smoke (as from a cigarette) coming from a point in the circuit or device. If any of this should ever happen, immediately remove power from the device or circuit by removing the line cord from the wall outlet. Don't use the device's on/off switch to shut off the power, as there might be dangerous current present. You must not restore power until the cause of the problem has been found and corrected.

Ground

Back in Chapter 5 we discussed how the term "ground" is used in reference to antennas. When we're talking about electrical circuits, ground means a point of *zero* voltage regardless of whether it's actually physically connected to an Earth ground. (The Earth itself is a point of zero voltage, of course!) However, the negative voltage terminal of a battery or power supply is often treated as the ground point; this is known as *negative ground*.

There are two types of schematic symbols for a ground connection. Part A of Figure 6-9 shows a symbol for a ground connection that is made directly to the Earth, such as might appear in an antenna circuit. Part B of the same figure shows the symbol for a *chassis* ground. This means the ground connections can be made directly to the metal chassis on which the circuit is built. This is how the symbols are used in the Novice and Technician written exams—part A is an Earth ground, part B is a chassis ground, and that is how you should answer any questions on the exam about them. However, if you have any experience with electronics, you have probably seen the symbol in part A used for chassis grounds, Earth grounds, and the negative terminal of a battery or power supply. In fact, you probably won't see the symbol in part B used much at all unless it's in a ham radio book or magazine. But be prepared to recognize it on the written exams.

The ground connection is a nice shorthand way of simplifying a schematic diagram. Look at part C of Figure 6-9. This is a "negative ground" circuit but without any ground symbol used. Now look at part D of the same figure. This is the same circuit as in part C, but with the ground symbol in part A used. Admittedly, this is not a great deal more simple than before. However, the reduction in complexity can be considerable when larger schematic diagrams are involved.

FIGURE 6-9:
Know the difference between the Earth and chassis ground symbols for the written exam. Using a chassis ground symbol to indicate a common connection point for different components can really simplify a circuit diagram.

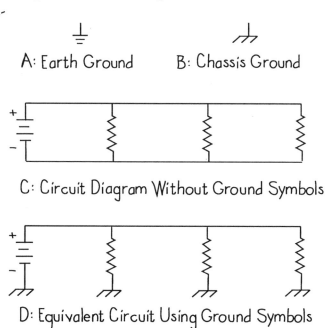

A: Earth Ground   B: Chassis Ground

C: Circuit Diagram Without Ground Symbols

D: Equivalent Circuit Using Ground Symbols

Alternating Current (AC)

Direct current flows in one direction only. But the electricity that flows into your home and is available at wall outlets is *alternating current* (AC). Alternating current is electricity that "takes turns" flowing in opposite directions. Alternating current is much more important than direct current in ham radio, since the output of your transmitter is a form of AC.

The AC that flows into your house from the electric company is in the form of sine waves, just like those we met back in Chapter 3 (take a look at Figures 3-3, 3-4, and 3-5 back there to refresh your memory). The frequency of the AC flowing in your house is 60 Hz. The positive and negative peaks of a sine wave represent the two directions of AC current flow. Like all sine waves, an AC wave starts at zero. It gradually builds to a peak in one direction (the 90° point) and then declines back to zero (the 180° point). The current then starts flowing in the opposite direction and increases to a peak (270°) and gradually decreases back to zero (360°). This happens to be the starting point (0°) of the next cycle, and so the AC cycle repeats.

Why is the current coming into your house AC instead of DC? The reason is because the most efficient way to produce the enormous quantities of electricity the world needs is through a device known as a *generator*, whose output is AC. A generator is based on the reciprocal relationship between electric current and magnetism we looked at a few pages ago. To produce electricity, a generator rapidly moves many turns of wire through a magnetic field produced by two or more magnets. As the turns of wire move through one of the magnetic fields, a current begins to flow in one direction, builds to a peak value, and then declines back to zero as the wire moves out of the magnetic field. As the conductor moves onward, it enters the second magnetic field. However, this time an electric current begins to flow in the opposite direction. Remember back in school when you learned that magnets have two "poles," a north pole and south pole? The second magnetic field is produced by a magnet of the opposite pole compared to the first. Since the "direction" of the magnetic field is opposite, so will be the direction of the current flow induced in the wire as it travels through the magnetic field. In real-world generators, the rotating coils of wire are mounted on shafts and the induced current flows through wires to one end of the shaft, where the current is taken from the generator at special contact points.

Capacitors

When we introduced the concept of electric fields earlier in this chapter, we used the example of two parallel copper plates, not connected in any way, which are charged with opposite polarities and thus have an electric field existing between them. This situation is shown in part A for Figure 6-10. This example wasn't one I picked out of the air—it happens to be a working model of a very common electronic component known as a *capacitor*, and the ability to store energy as an electric field (or "electrostatically") is known as *capacitance*. A capacitor consists of two parallel conductors separated by an insulating material known as a *dielectric*. The dielectric can be almost any insulator, such as mica, paper, or air. The schematic symbol for a capacitor of a single fixed value is shown in part B of Figure 6-10, while the schematic symbol for a capacitor whose value can be varied (much like a potentiometer) is shown in part C. Finally, part D of Figure 6-10 shows the schematic symbol for a special type of capacitor known as an *electrolytic* capacitor.

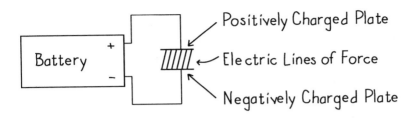

Positively Charged Plate

Electric Lines of Force

Negatively Charged Plate

A: Model of a Capacitor

B: Fixed Capacitor Symbol   C: Variable Capacitor Symbol

FIGURE 6-10:
Again, note carefully
where the + always
is on the symbol for
an electrolytic
capacitor.

+ ⟵ This Side Must Always Be
Connected to Positive Voltage Source

D: Electrolytic Capacitor Symbol

Capacitors are often described in terms of the dielectric used. For example, a *paper capacitor* is one in which the dielectric is paper. By the same token, a *mica capacitor* is one that uses thin sheets of mica as the dielectric. The type of dielectric used in a capacitor influences its capacitance. Other factors that influence how much charge a capacitor can store are the spacing between the plates, the area of the plates, and the number of plates. (Most variable capacitors—the schematic symbol shown in part C of Figure 6-10—have multiple plates.) If the spacing between the plates increases, the capacitance *decreases*. If the area of the plates increases, then the capacitance also *increases*.

Electrolytic capacitors use a special chemical paste as a dielectric. When electricity is first applied to an electrolytic capacitor, the paste is "activated" and spreads to form the dielectric on one set of plates. Electrolytics are widely used because they can pack a lot of capacitance in a physically small area. Most electrolytics are "polarized" meaning the "positive" side of the electrolytic capacitor must be connected to a positive voltage source. The positive side of the electrolytic is indicated by a +, as shown in part D of Figure 6-10. The + symbol is always adjacent to the straight line in the capacitor symbol, not the curved line.

Capacitance is measured in *farads*. However, this is too big a unit for capacitors used in practical electronics and radio work. Instead, *microfarads* (abbreviated μF, or 0.000001 farad) and *picofarads* (pF, or 0.000000000001 farad) are used instead. These are really tiny units, with one farad being equal to 1,000,000 μF and 1,000,000,000,000 pF.

Another important measurement for capacitors is *voltage rating* or *working voltage*. This describes the maximum voltage that can be applied to a capacitor without destroying it. If the voltage rating is exceeded, the dielectric will break down and the capacitor can literally "explode"! Most capacitors will have their capacitance (in either μF or pF) and voltage rating printed on their casings. To provide a good margin of safety in case of momentary voltage surges or "spikes," it's a good idea to use capacitors with voltage ratings about 50% higher than the highest voltage normally found in a circuit.

In a DC circuit, current flows into a capacitor until it reaches the voltage and capacitance limits of the capacitor. Once the capacitor is fully "charged," no more current can flow into a capacitor and it is, for all practical purposes, like an open circuit. But in an AC circuit, a capacitor will alternately be charged and discharged as the current flow reverses direction. Thus, AC can flow "through" a capacitor while DC cannot. However, a capacitor opposes any change in the voltage applied to it, and this opposition is known as *capacitive reactance*.

Reactance is similar to resistance (it's also measured in ohms) but—unlike resistance—the reactance of any capacitor *decreases* as the frequency of the AC flowing through it increases. Because of this capacitive reactance, capacitors are often used to "bypass" high frequency AC signals to ground.

A few pages ago we saw how to determine the total resistance of resistors connected in series and parallel. To determine the total capacitance of capacitors connected in series and parallel, the formulas we use are the opposite of those used for resistance! In brief:

To Find the Total Capacitance of Capacitors in Series

Add the reciprocal of the value of each capacitor take the inverse of the sum. If we have three capacitors in series, the formula to use would be:

$$\frac{1}{\left(\frac{1}{C1}\right) + \left(\frac{1}{C2}\right) + \left(\frac{1}{C3}\right)} = \text{total capacitance of three capacitors in series}$$

The total capacitance of any number of capacitors connected in series will always be less than than the smallest value capacitor that is part of the series connection.

To Find the Total Capacitance of Two Capacitors in Series

This is a variation on the formula used to find the value of two resistors in parallel. The formula is:

$$\frac{(C1 \times C2)}{(C1 + C2)} = \text{total capacitance of two capacitors in series}$$

To Find the Total Capacitance of Capacitors in Parallel

Again, do the opposite of what you would do with resistors! All you have to do is add the values of each capacitor. Suppose that we have three capacitors in parallel:

C1 + C2 + C3 = total capacitance of three capacitors in parallel

Inductors

Capacitors store energy in electrostatic fields. By contrast, *inductors* store energy in magnetic fields. Inductors are basically just coils of wire. As electric current flows through an inductor, a magnetic field develops around the inductor. Winding the wire into a coil helps concentrate the magnetic lines of force into a smaller area. The ability to store energy in a magnetic field is called inductance, and it's measured in *henrys*. Like the farad, the henry is too large for most practical electronics and radio applications. Instead, the *millihenry* (one-thousandth of a henry, abbreviated mH) and *microhenry* (one-millionth of a henry, abbreviated µH) are used. The operating principles of an inductor depend on the relationship between magnetism and electricity that we looked at earlier in this chapter. As current flows through the coiled wire of an inductor, a magnetic field is set up around the inductor. When AC flows through an inductor, the magnetic field expands and contracts as the AC changes direction and strength. The changing magnetic field moves across the coiled wire making up the inductor, and sets up a voltage opposite in polarity to the voltage applied to the inductor. This "counter-electromotive force" means that inductors will oppose the flow of current through them. This is known as *inductive reactance*, and like capacitive reactance and resistance is measured in ohms. Inductive reactance is directly proportional to frequency; the higher the frequency of the AC flowing in an inductor, the higher its reactance (opposition to current flow).

There are two basic parts to any inductor: the coil of wire itself and the *core*. The core of the coil could be air (and this is frequently the case) or a form made from different materials. Sometimes the core is an electrically and magnetically neutral material such as plastic. In other cases, the core is a material such as iron. An iron core helps contain more of the magnetic lines of force within the inductor, and *increases* the inductance. If you were to take an air core inductor and slip an iron core inside it, the inductance would immediately jump significantly. The term *permeability* is used to describe how a core material affects the inductance of a coil of wire. Air has a permeability of 1, while materials such as iron have permeabilities well into the hundreds. Some core materials actually reduce the inductance of a coil; for example, adding a brass core to a coil would produce a lower inductance. The factors that influence the inductance of an inductor are the core material, the diameter of the core, the length of the wire coil, and the number of turns of wire used to "wind" the coil.

Figure 6-11 shows the schematic symbols used to represent inductors. Part A shows an air core inductor, part B is the symbol for an inductor with an iron core, and part C shows an inductor whose inductance can be varied. By the way, the shape of the core over which the coil is wound can vary. For example, the core might be cylindrical or a doughnut-like "toroid" shape. Regardless of the shape of the core, the same schematic symbols are used.

When inductors are connected in series and parallel, the total inductance is calculated using the same formulas used when resistors are connected together. For your exam, memorize the formulas for calculating the values of resistors connected in series and parallel. The same formulas can be used for inductors, and then do the *opposite* if the question involves capacitors connected in series and parallel.

A: Air Core Inductor

B: Iron Core Inductor

C: Variable Inductor

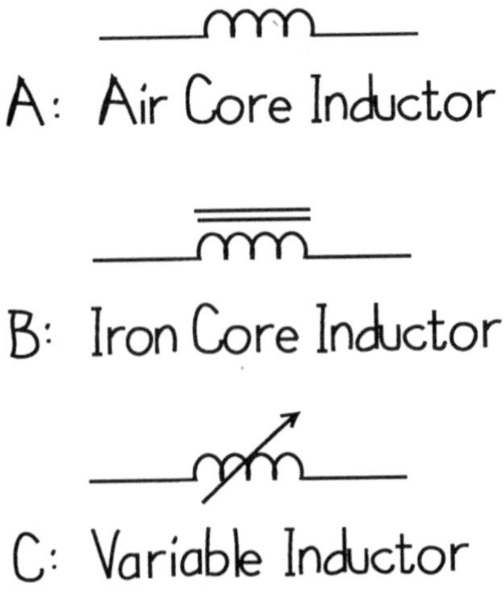

FIGURE 6-11: Memorize these inductor symbols—you might see them again on the exam.

Shielding

A *shield* is made from a conductor connected to ground. As its name implies, a shield is designed to contain electromagnetic energy within the area enclosed by the shield so that it does not affect components adjacent to it. If you have a microwave oven in your house, look carefully at the glass covering of its front door. You'll usually see a fine metallic mesh underneath the glass. That mesh absorbs microwave energy that could otherwise leak out through the door.

Inside any transceiver, receiver, or transmitter are several sources of electromagnetic energy and varying magnetic fields. These can produce stray currents in adjacent conductors and unintentionally "couple" separate circuit sections together through mutual induction. To prevent this, shielding around sources of particularly strong fields can be used to isolate those components or circuits from the rest. If you ever work on a transmitter, you must be very careful about operating the transmitter with some of the shielding removed. If the transmitter operates on the HF bands, the result could be radiation that could cause harmful interference to other stations. If the transmitter (or amplifier) operates at VHF or UHF, the resulting stray radiation could cause injury to body tissues as well as QRM. In the next chapter, we'll see how VHF and UHF radio waves can be harmful at high power levels.

The Effects of Capacitive and Inductive Reactance

It's easy to get capacitive and inductive reactance confused, so let's summarize the important facts about both:

- Capacitive reactance opposes any change in the voltage applied to a capacitor.
- Capacitive reactance decreases as the frequency of the AC increases.
- Inductive reactance opposes any change in the current applied to an inductor.
- Inductive reactance increases as the frequency of the AC increases.

So what's the big deal about capacitive and inductive reactance? They're important because of how they can affect alternating current flow. So far, we've assumed that the current and voltage levels of an AC wave are always the same; that is, when the voltage is at its positive and negative peaks, so is the current. However, capacitive and inductive reactance in an AC circuit can throw this neat arrangement out of joint. It's possible for the voltage peaks in an AC wave to occur before the current peaks, for example, and in such a case we'd say the voltage "leads" the current. (Another way to put it would be to say the current "lags" the voltage.) It's also possible for the opposite to take place, and for the current to "lead" the voltage.

Last chapter, we defined the correct electrical length of an antenna as that length that would allow the radio energy applied to the antenna to travel from one end of the antenna and back within one cycle. We also mentioned that a loading coil (an inductance) can make an antenna electrically "longer" and capable of operation on lower frequencies than otherwise would be the case. This is because the inductance of the loading coil changes the amount of time it takes each cycle to travel to one end of the antenna and back, thereby effectively making a short antenna perform like a longer one. We also mentioned that short antennas with loading coils have a narrower bandwidth over which they're resonant than longer antennas of the full physical size for a given frequency range. This is because the inductance of the loading coil changes as the frequency changes.

Capacitive and inductive reactance are used together in devices known as *antenna tuners* or *antenna tuning units*. (They're also sometimes called *transmatches*.) An antenna tuner contains variable capacitors and inductors so that a wide range of different capacitive and inductive reactances can be created. The antenna tuner is placed between your transmitter and antenna, and allows you to precisely "match" your transmitter and antenna for any frequency you're operating on. This lets you get as much power as possible from your transmitter into your antenna, and also lets you use a single antenna over a wide range of frequencies.

How can an antenna tuner do this? Because the antenna tuner can match the *impedances* of your transmitter and antennas.

Impedance is the combined effects of capacitive reactance, inductive reactance, and ordinary resistance. Impedance is also measured in ohms. Impedance plays a big role in whether a transmitter can be properly loaded into an antenna (remember that term from the last chapter?). The impedance of most transmitters at their antenna terminal is 50Ω, and this impedance is constant regardless of operating frequency. The impedance of antennas varies by type and whether or not the antenna is resonant at the transmitter's frequency. For a resonant dipole, the impedance at the center point where energy from the transmitter arrives is approximately 75Ω while for a resonant vertical the input impedance is about 35Ω. Maximum power can be transferred between two circuits (such as from a transmitter to an antenna) when their impedances are identical. The slight differences in impedances between the two types of antennas and the typical 50Ω transmitter impedance is not significant, and so 50Ω represents a good compromise for the transmitter output.

What happens if the impedances are wildly mismatched, as when the transmitter output is 50Ω and the antenna impedance is in the hundreds of ohms? The results are a lot like what happens when someone who only speaks English tries to communicate with someone who only speaks Japanese—a little "communication" can maybe take place, but a whole lot is lost! If you have a severe impedance mismatch, as little as just one watt out of every ten that comes out of your transmitter would actually be radiated by your antenna. The rest would be reflected back from the antenna and into your transmitter. Not only does this waste transmitter power, but it can also damage your transmitter. Impedance mismatches this bad are easy to produce if you try to use an antenna for a band it's not resonant on (such as trying to use a 40-meter band dipole on 20 meters). The solution is to use an antenna tuner to alter the antenna system impedance so that your transmitter always "sees" a 50Ω impedance. An antenna tuner lets you use a single band antenna on multiple bands and also to achieve a perfect impedance match if you're using a dipole or vertical antenna of the right length to be resonant on the band on which you're transmitting.

An antenna tuning unit sounds pretty terrific, doesn't it? But, like the loading coil technique mentioned in the last chapter, it has some real limitations. You can use an antenna tuner to operate 20 meters using a 40-meter band dipole, but the performance won't be anywhere near as good as it would be if you were using a dipole of the right physical length for 20 meters. Almost all of your transmitter's energy will reach the antenna and be radiated, but it won't be radiated as effectively as if you were using an antenna of the proper physical size. Antenna tuners can't work miracles, but they can let you operate on frequencies you normally couldn't and also help to squeeze every last bit of performance out of any antenna system.

All this explains why inductive and capacitive reactance is important. They can alter when the voltage and current peaks occur in the energy from your transmitter, and that in turn can change the impedance your transmitter sees from the antenna. When the transmitter and antenna impedances are the same (or nearly so), most of the energy from your transmitter gets radiated into space.

Switches

A *switch* is a mechanical device that lets you control whether or not current flows in a circuit. You're already familiar with the concept of a switch and how it works (unless you never turn on the lights in your house after dark), but switches come in several different types. You'll need to be familiar with these.

Part A of Figure 6-12 shows the simplest type of switch, the *single-pole, single-throw* (SPST). This is also call an "on/off" switch. There's only one path the current can take; current can't flow if the switch is open (as it is in part A) but current will flow if the switch is closed. Part B shows a more complex type of switch in which current can flow in one of two different directions depending on the setting of the switch. This is called a *single-pole, double-throw* (SPDT) switch. A common application for this type of switch is selecting between two different antennas.

We sometimes need to switch two current paths (that is, wires) at the same time. Part C shows the schematic symbol for a *double-pole, double-throw* (DPDT) switch. A DPDT switch is like two SPST switches that are closed or opened simultaneously (the dotted line indicates the two contacts move together, not separately).

Finally, sometimes we need what might be called a "single-pole, multiple-throw" switch. One common example might be a switch to choose between several different antennas, or perhaps a switch on a transceiver to choose between different ham bands. Part D of Figure 6-11 shows a single-pole switch which can be moved to five different positions.

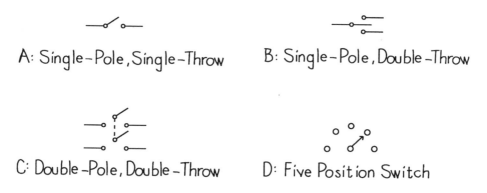

A: Single-Pole, Single-Throw   B: Single-Pole, Double-Throw

C: Double-Pole, Double-Throw   D: Five Position Switch

FIGURE 6-12: Schematic symbols for different types of switches.

Schematic Symbols for Other Components

You'll need to be able to identify the schematic symbols for other electronics components, although (at the time this book was written) you won't have to understand how these components work. You'll need to know more about their operation if you go for a General or higher class license, and you'll eventually want to know about these devices as you get more experienced in ham radio. But to get a Novice or Technician license you only have to recognize these symbols.

The two most common transistors are known as the *PNP* and *NPN* types. Figure 6-13 shows the schematic symbol for a PNP transistor at part A and the symbol for a NPN transistor at part B. They look very similar, but the key to telling them apart is whether the arrow is pointing in or out of the circle. If the arrow is Not Pointing iN, the transistor is a NPN type. If the arrow is pointing in, the transistor is a PNP type.

Vacuum tubes are not often used in today's ham equipment, except to amplify radio signals at high power levels (such as at 1000 watts). Part C of Figure 6-13 shows the schematic for a tube known as the *triode*. A triode has three parts, known as the plate, grid, and cathode. You might think you see four parts to the triode, but the cathode includes a "heater" element that isn't considered separate from the cathode (even though it looks that way in the schematic symbol).

You'll need to be able to recognize the schematic symbol for an antenna. We've already used it earlier in this book, so it should be familiar to you. The "official" (that is, one you'll see on the exam) symbol is shown in part D. (It even looks a bit like an antenna, doesn't it?) However, you'll often see an antenna symbol like that in part E in ham radio magazines and equipment manuals.

While the current Novice and Technician exams don't ask you about *diodes*, all real hams should know what a diode is and the schematic symbol for one. A diode lets current flow in one direction (called the *forward direction*) while blocking it in the opposite (or *reverse*) direction. The schematic symbol for a diode is shown at part F of Figure 6-13. Diodes are widely used to convert AC to DC. They do so by permitting current to flow during one half of the AC cycle but not the other. A common circuit in power supplies is to connect four diodes in a square so that two of the diodes conduct during one half of an AC cycle while the other two diodes conduct during the other half. This produces a pulsating DC output which is changing in value but flowing in only one direction. Some diodes glow when electricity is applied to them. These are called *light emitting diodes* (LEDs), and are common in ham gear as well as consumer electronics devices.

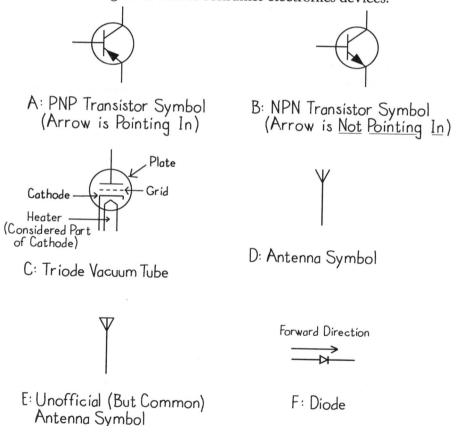

A: PNP Transistor Symbol
(Arrow is Pointing In)

B: NPN Transistor Symbol
(Arrow is <u>Not</u> <u>Pointing In</u>)

Plate
Grid
Cathode
Heater
(Considered Part
of Cathode)

C: Triode Vacuum Tube

D: Antenna Symbol

E: Unofficial (But Common)
Antenna Symbol

Forward Direction

F: Diode

FIGURE 6-13: These are some of the most commonly seen schematic symbols in ham radio.

Amplifiers and Oscillators

At the time this book was being written, there were no questions on either the Novice or Technician exams about amplifiers or oscillators. However, I don't think you can be a real ham without knowing a little something about these two subjects, so let's dig in.

We briefly mentioned amplifiers back in Chapter 2. An *amplifier* is a circuit that takes an input signal, such as the output of a transmitter, and produces an output signal that is a much stronger version of the input signal. For example, most HF ham transceivers today have an output power of about 100 watts, well below the 1500 watts of power permitted General and higher class hams on the HF bands. An external power amplifier can take the transceiver output as its "drive" signal and produce an output that is a reproduction of the signal from the transceiver, except that it has a power level of 1500 watts instead of 100.

Some amplifiers are described as *linear* amplifiers, meaning that the output signal of the amplifier is directly proportional to the input signal driving it. (The term "linear" means that if you were to plot the input and output signals on a graph, the result would be a straight line.) Amplifiers come in different "classes," and these classes differ how in distorted the output signal is compared to the drive signal. Class A amplifiers have the least output distortion of all, and are used in such applications as stereo systems where we want the output to be as accurate a reproduction of the input as possible. However, class A amplifiers are not very efficient, and produce less output power for a given amount of current and voltage than other classes. Class B amplifiers are more efficient, but produce more distortion than class A amplifiers. Class B amplifiers are often used to amplify SSB signals. Class C amplifiers are even more efficient than class B, but produce still more distortion. In fact, class C amplifiers can't be used to amplify SSB signals, since they distort the amplitude of the carrier so badly that human speech can't be understood. However, class C amplifiers can be used with FM and CW signals since carrier amplitude variations aren't important with those modes. Just remember that an amplifier suitable for amplifying an FM signal may not be suitable for amplifying SSB. Fortunately, most amplifiers are clearly described as being "multimode," "FM/SSB," "FM only," etc.

All transmitters are *oscillators*. Oscillators are circuits that continuously generate high frequency AC sine waves. The frequency of an oscillator's output signal is determined by a *tuned circuit*. The most basic type of tuned circuit consists of a capacitor and inductor connected in parallel. Energy applied to the tuned circuit will "bounce" back and forth between the capacitor and inductor, setting up oscillations. To keep the oscillations going, some of the output energy is fed back into the tuned circuit; this process of sending some of the output energy back into the input is known as *feedback*. Feedback can be accidental and detrimental. For example, you may have heard a high pitched "squeal" sometimes over a public address system. This is an example of feedback, although an unwanted one.

Not all tuned circuits are made from inductors and capacitors. For example, quartz crystals are often used. These *crystal oscillator* circuits have superb frequency stability and were very popular for many years; you had to "plug in" a new crystal whenever you wanted to change the oscillator (or transmitter) frequency. (When I got my Novice license many years ago, FCC rules required all Novice transmitters to use "crystal control." A dozen or so crystals were always within arm's reach when I was on the air!) Most oscillators used in ham transmitters today are designed so their frequency can be altered over a wide range. This type is called a *variable frequency oscillator* (VFO), and lets you pick any frequency in a ham band you want (according to your license class, of course!) to transmit on.

Block Diagrams of Common Circuits

Don't get too freaked by the schematic symbols we introduced in this chapter. You won't have to interpret complete circuits composed of schematic symbols on your exam. However, you will need to be familiar with block diagrams of some common items of ham equipment. You'll probably be asked to identify an unlabeled block on a couple of these or similar diagrams. Moreover, it's good to have a general idea of what's going on inside your equipment anyway.

The simplest type of transmitter, one for CW, is shown in Figure 6-14. The RF signal is first generated in an oscillator stage. The output frequency of the oscillator can be controlled either by a quartz crystal or some sort of VFO. However, the output of the oscillator needs to be amplified before it is transmitted. We also need to isolate the oscillator from the antenna system, since an oscillator needs to have a constant output impedance for best frequency stability. The solution is to use a *buffer* or *driver* amplifier to amplify the oscillator's output and provide a constant impedance at the oscillator's output. The output of the buffer/driver is fed to a *power amplifier*, which boosts the signal to the desired level power before sending it to the antenna. The *code key* is located between the buffer/driver and power amplifier stages, and is used to form the Morse code characters.

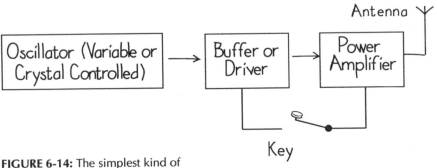

FIGURE 6-14: The simplest kind of radio transmitter—one used for CW.

To receive signals from such a transmitter, a receiver like the one in Figure 6-15 is used. This type of receiver is called a *superheterodyne*, and involves taking a signal received on one frequency, such as 7230 kHz, and converting it to a fixed *intermediate frequency* (IF) such as 455 kHz. A superheterodyne converts all signals it receives to the intermediate frequency since it is easier to build a receiver to amplify and demodulate signals at one frequency instead of a broad range. The signal from the antenna is first applied to a *radio frequency* (RF) *amplifier* stage, which boosts the strength of the receiver signal so it's strong enough for the rest of the receiver circuitry to process it. The RF amplifier output goes to a *mixer* stage, where it is combined with a signal from a *local oscillator*. As its name indicates, the mixer combines the received and local oscillator signals together and produces an output which is always equal to the intermediate frequency. The mixer output is sent to the intermediate frequency amplifier, which boosts the strength of the IF signal. The IF amplifier output goes to the *detector* stage, which is where the audio or other intelligence is extracted (demodulated) from the IF signal. If the received signal is CW or SSB, a *beat frequency oscillator* (BFO) provides a replacement carrier so such signals can be demodulated. The output of the detector is an audio signal, which is then sent to an audio amplifier stage that drives the speaker in your receiver. This kind of superheterodyne receiver can be used to receive AM, SSB, CW, and RTTY signals.

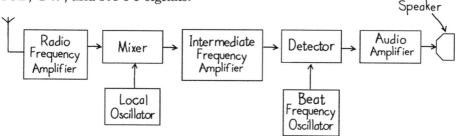

FIGURE 6-15: This type of superheterodyne receiver can be used to receive AM, SSB, and CW signals.

For FM systems, the transmitter and receiver circuits are different. Figure 6-16 shows a transmitter used for FM. When you speak into the microphone, you set up an AC voltage that varies according to your voice. The signal from the microphone is boosted by an audio amplifier and sent to a *clipper* stage. The clipper stage limits the audio signal to amplitudes that will not cause overmodulation and excessive deviation. The clipper output goes to a *reactance modulator*, where it is combined with the output of an oscillator. The reactance modulator will vary the frequency of the oscillator's output signal in accordance with the variations in the audio signal. Back in Chapter 3, we mentioned that it's easier to frequency-modulate a signal at a lower frequency and then increase, or "multiply," the signal in frequency until the desired operating frequency is reached. This is why the output of the reactance modulator is fed to a series of *frequency multipliers* until the desired transmitting frequency is reached. When it is, the signal is fed to a power amplifier and boosted to the desired level before transmitting.

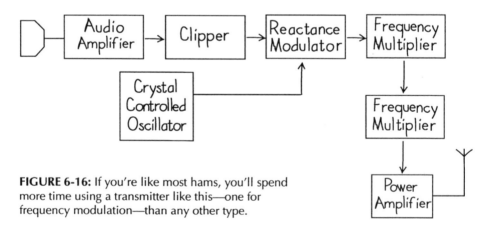

FIGURE 6-16: If you're like most hams, you'll spend more time using a transmitter like this—one for frequency modulation—than any other type.

FM receivers are variations of the superheterodyne receiver shown in Figure 6-15, but with two new circuits replacing the standard detector and BFO stages. Figure 6-17 shows a typical

FM receiver. The RF amplifier, local oscillator, mixer, and IF amplifier circuits are the same as before. However, note that a *wide filter* circuit has been placed between the mixer and IF amplifier stages. This filter is equal to approximately the total bandwidth of the desired FM signal. The output of the IF amplifier goes to a *limiter* stage. The limiter removes any remaining amplitude modulated components (such as noise) from the signal, so the result is one varying only in frequency. This "scrubbed" signal is then sent to a *frequency discriminator*, which produces an audio signal in accordance with the frequency variations of the signal supplied to it. The audio output signal from the frequency discriminator is then boosted by an audio amplifier and fed to a speaker.

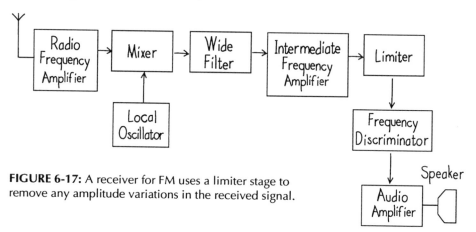

FIGURE 6-17: A receiver for FM uses a limiter stage to remove any amplitude variations in the received signal.

Now Relax. . .

This chapter has probably been the most difficult one in the the book so far. Don't worry if you didn't get it all the first time you read though it. Go back and re-read this chapter from time to time—and read it *slowly*. You can't read technical material as rapidly or easily as a novel; you have to be more actively involved in trying to understand it. (Well, maybe you can read technical material as quickly as a novel. I sure can't, and neither can most other people!) So if you're a little confused about what you've just read, don't panic. Just try again—and again—and eventually it will all come together for you.

And I strongly advise you to continue learning more about electronics and radio even if you don't plan on getting a General or higher class license. Ham radio's more fun when you understand what's going on inside those boxes on your desk!

The Decibel: Threat or Menace???

The *decibel* (dB) is a way to describe the ratio between two power levels. It's named after Alexander Graham Bell and reflects the human ear's response to an increase in volume (i.e., a change in power level). There's a nice, complicated formula for computing the decibel difference between two power levels, but you don't need to know it for your Novice or Technician exam. But every ham should understand how decibels work and their practical impact on ham radio. You see, sometimes I think the decibel is inherently dangerous. Why? Because I don't think any technical term is so widely misunderstood or has been deliberately misused more often to mislead people.

The confusion starts because the decibel is *logarithmic* rather than a linear, arithmetical term. Say you have a transmitter putting out 100 watts and decide to boost its power to 200 watts. Would the transmitter sound twice as loud? Common sense and the linear, arithmetical way of thinking would tell you yes, but the answer is *NO*. Doubling the transmitter power will produce only a slight, barely noticeable change in the strength of the signal heard from a receiver. You would have to increase the transmitter power *ten times*—to 1000 watts—before the transmitted signal would sound twice as loud! That's what logarithmic response is all about.

Decibels can be used to express a loss as well as a gain in power. Here's a short list of some common decibel values and the multiple of gain or loss that dB represents:

3 dB = gain or loss of 2

6 dB = gain or loss of 4

10 dB = gain or loss of 10

20 dB = gain or loss of 100

It's important to remember that the decibel is the ratio between two power levels. You'd never know this by some claims appearing in advertisements for antennas and power amplifiers, however. You'll see ads boasting some antenna has "6 dB" gain. Your first reaction might be "Hey! That's terrific! I want one too!" But what is that 6 dB gain compared to? Is it a half-wave dipole or a piece of wet spaghetti??? The decibel isn't an absolute, independent measure like a kHz or meter (for length); if you don't know what the beginning power level is, than any sort of dB "gain" or "loss" is meaningless junk. This kind of stuff goes on more often than you might expect. For example, I know of one case where a new antenna for hand-held two-meter FM transceivers was described as having 6 dB gain. Very impressive—but the ad didn't mention that the gain was computed in comparison to a shortened flexible "rubber duck" antenna, which has a substantial power *loss* compared to a half-wave dipole! (Let's recognize the ARRL and *QST* for refusing to run ads for antennas and related items which claim a certain amount of "dB gain." Some other ham magazines should follow their example!)

For everyday hamming, keep in mind that doubling your transmitter power won't make your signal sound twice as loud, nor will even quadrupling your power do the trick! But look on the bright side—even if you're using modest transmitter power, you're not that much less loud than stations who are using all the FCC allows. Keep in mind that small gains (or losses) in dB, such as 3 dB or less, really aren't that significant. (Come to think of it, a 6 dB gain or loss isn't really that important most of the time.) Take claims of so much gain in dB for various antennas and other items with a healthy dose of salt, and maybe we can restore the decibel to its rightful place as a useful engineering measurement instead of a tricky way to deceive those who are less technically savvy than they should be.

CHAPTER SEVEN

Beam Antennas, Feedlines, Test and Measurement Gear, and Safety

I F YOU'VE MADE IT THIS FAR in the book, congratulations—you're now on the home stretch for your ham ticket! In this chapter, we'll wrap up the remaining technical items you'll need to know for your exam. Unlike the last chapter, the information in this chapter is practical stuff you'll use on a regular basis to get the best performance from your station. We'll also look at some important safety guidelines so you'll be enjoying ham radio for many years to come!

Beam Antennas

Back in Chapter 2, we mentioned that some antennas are directional, meaning they radiate more of your transmitter's energy in certain directions than others and also receive signals better from those directions than others. As a group, directional antennas are called *beams* because they "beam" your signals in the direction you want.

Regardless of their design, all beam antennas work by adding additional metallic antenna parts, called *elements*, to the one part (such as a half-wave dipole) that actually receives energy from the transmitter. The element receiving energy from the transmitter is called the *driven* element. The other elements, known as *parasitic* elements, of the beam antenna receive energy from the driven element. When the driven element radiates the electromagnetic energy from the transmitter, currents are set up in the parasitic elements. As currents flow in the parasitic elements, electromagnetic fields are set up around them as well. The electromagnetic fields around the parasitic elements begin to affect the electromagnetic field around the driven element. The result is that the electromagnetic field around the beam antenna gets distorted. The field is much stronger in one direction (known as a *major lobe*) than in other directions.

The most widely used beam antenna is the *Yagi*, named after its Japanese designer. Figure 7-1 shows a basic three-element Yagi antenna. The driven element in a Yagi is a dipole that is resonant near the middle of the band (such as 20 meters) that the Yagi is designed to operate on. Besides the driven element, the Yagi has two elements called the *reflector* and the *director*. The reflector is approximately 5% longer than the driven element while the director is approximately 5% shorter. All three elements are mounted in parallel to each other on a support structure known as a *boom*. The major lobe (also known as *forward lobe*) of a Yagi antenna is in the direction the director faces, as shown in Figure 7-1. The driven element will always be between the director and reflector. While this simple Yagi only has one director, it's possible for a Yagi to have more than one director. These extra directors "focus" the forward lobe of the Yagi into a narrower, more concentrated field.

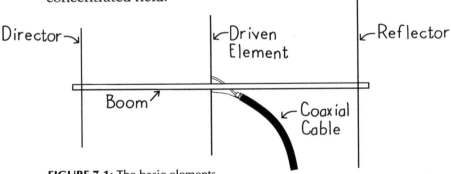

FIGURE 7-1: The basic elements making up a Yagi antenna.

Since the driven element of a Yagi is a dipole, the dimensions of the Yagi elements are about the same as those of a half-wave dipole according to the formula given back in Chapter 5. As a result, Yagis for 80 meters are very rare and only slightly more common on 40 meters due to the large amounts of space required. (Elements of a Yagi can be "loaded" to make them resonant on bands they are physically short for, but this degrades the Yagi's performance.) Yagis are most common on 20 meters and higher, and Yagis for VHF and UHF work are no bigger than most outdoor television antennas.

The polarization of a Yagi depends on how it is mounted. If it is installed so that its elements are perpendicular (or vertical, as shown in Figure 7-1) with respect to the Earth's surface, the radio waves from it will be vertically polarized. The reverse is also true—if a Yagi is installed so that its elements are parallel to the Earth's surface (horizontally), the radio waves from it will be horizontally polarized.

Running a close second to the Yagi in beam antenna popularity is the *cubical quad*, as shown in Figure 7-2. The basic "quad" consists of two parallel four-sided wire "boxes" called *loops*. The wire making up each loop is approximately one wavelength long for the band the quad is designed to operate on, and the wire is folded so that each side of the loop is about a quarter wavelength long. Both loops are mounted on a boom, and the separation between the loops on the boom is approximately a quarter wavelength at the middle point of the ham band the quad is designed for. One of the loops, again called the driven element, receives power from the transmitter. The other element is again the reflector, and the wire making up it is a little bit longer than that of the driven element. Director loops, which will be slightly shorter than the driven element, can be added in front of the driven element.

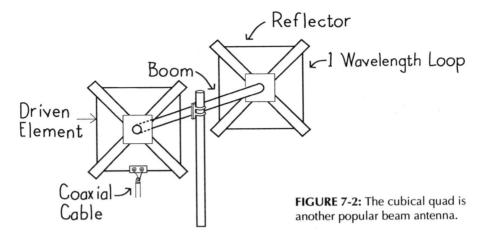

Reflector

Boom

Driven Element

Coaxial Cable

—1 Wavelength Loop

FIGURE 7-2: The cubical quad is another popular beam antenna.

Electromagnetic energy from the transmitter is usually applied either to the quad at the bottom side of the driven element loop (as shown in Figure 7-2) or to one of the two vertical sides of the driven element loop. Where the driven element receives the power from the transmitter determines the polarization of a cubical quad antenna. If the quad is "fed" at the bottom as in Figure 7-2, the polarization of the antenna is horizontal (this is because that side of the quad is parallel to the Earth's surface). If the quad is fed at one of the vertical sides of the driven elements, its polarization is then vertical. Some quads have their loops rotated so that all sides are at a 45° angle relative to the Earth (the elements have a "diamond" shape). If such a quad is fed at the bottom corner of the driven element, the polarization of its signals will be horizontal. If the feed point is one of the sides, the polarization will be vertical.

Finally, there's a variation of the cubical quad known as the *delta loop* antenna. The delta loop has elements that are triangular, consisting of a one wavelength-long length of wire folded into three sides. All the rules we've just mentioned about cubical quads also apply to delta loops.

Both Yagis and quads come in *monoband* and *multiband* versions. A monoband beam is one that will work on just one ham band, such as 20 meters. A multiband beam works on several different bands, with 20, 15, and 10 meters being common. (In fact, a beam for 20, 15, and 10 is often called a *tribander*.) A multiband Yagi uses different loading coils (often called *traps*) in each element to make them resonant on different bands, while a multiband quad has multiple "nested" one-wavelength loops in each element. The loop for the lowest frequency band (which would be the longest wire) is the outermost loop of an element while the loop for the highest frequency band (the shortest wire) is the innermost loop. Thus, a triband quad would have three wire loops in each element.

Regardless of type, all beam antennas should be installed as high above the ground as possible. Due to their size, beams for below 30 MHz are usually installed on rigid metal support structures known as *towers*. Towers may be *self-supporting* types, meaning they can be installed directly into the ground without any other attachment to other structures or supporting hardware. Towers for larger antennas usually require the use of so-called *guy wires*. Guy wires are sets of cables from the tower to ground, with each of the wires connected to a stake driven into the ground. These wires help support the tower and balance out the effects of wind, much like ropes staked to the ground help balance and support a tent. Beams for frequencies above 30 MHz are smaller, and can be installed on support poles intended for television antennas or directly on a roof. You'll also need some way to turn a beam so that its forward lobe is pointing in a desired direction. A *rotor*, similar to those used with outdoor television antennas, does this. For HF antennas, so-called "heavy duty" rotors capable of moving heavy loads are necessary. For VHF and UHF beams, ordinary TV antenna rotors can often be used.

If you are able to install some sort of beam antenna at your station, do so! There's no single change you can make to improve the performance and communications range of your station that will have as big an impact as changing from a dipole or longwire antenna to some sort of beam. Not only will your signals sound louder to stations in the direction of your forward lobe, but interference from stations located to the sides and back of your forward lobe will be greatly reduced.

The Great Debate: Yagi Versus Quad

If you want to start a good fight among some hams, just ask for opinions about which type of beam antenna is best: a quad or Yagi. Let's say the question is something like this: "Which is better, a three-element Yagi or a three-element quad, assuming both are at the same height?" The arguments that follow often sound more like the rantings of religious fanatics instead of engineers and scientists!

The scientific and engineering data on this matter is a draw (although I bet some quad and Yagi enthusiasts out there will even dispute this statement!). However, the performance differences found in tests are generally so minor—and often contradicted by other studies—that most professional engineers who have studied the question agree there is no reliably measurable difference in performance between quads and Yagis with the same number of elements installed at the same height. However, a considerable amount of folk wisdom (some of it even approximately true!) has built up in the ham community about the relative merits of quads and Yagis. The consensus—such as it is—is that Yagis with four or five elements are superior to comparable quads so long as they are mounted at high elevations above the ground (say 60 feet or more). If you're limited to three elements or fewer and/or must install your antenna at a lower elevation above the ground, then a quad generally offers better performance. This is particularly true if you're comparing a multiband Yagi to a multiband quad.

However, there are some other considerations in choosing which beam is best for you. Yagis are invariably more rugged antennas than quads, and can withstand high winds and ice far better. (A quad the morning after a storm with freezing rain is not a pretty sight!) Yagis installed for horizontal polarization take up less space than a comparable quad; if the elements are loaded and therefore shortened, a Yagi can fit into an area too small for a quad. However, quads are generally less expensive and lighter. An ordinary TV rotor can be used to turn quads for such bands as 15 and 10 meters, and the supporting towers and other hardware don't have to be quite as heavy duty as for a Yagi. And quads are relatively easy to construct compared to Yagis.

Okay, I know I'm supposed to offer some suggestions, so here goes: If you can get a beam high into the air and will be using more than three elements, go for a Yagi. If you live in an area with adverse weather, particularly snow and ice, and want to stay on the air all winter, go for a Yagi. If you can't get your antenna very high into the sky and are restricted to three elements or less, go for a quad. If you're on a tight budget, go for a quad. If you're a Yagi or a quad fanatic that I've managed to offend, I'm sorry.

Feedlines

In Chapters 2 and 5, we used the term *feedline* without going into too much detail about what we mean when we use the term. A feedline is the conductor(s) used to get radio energy from your transmitter to your antenna; it's also called a *transmission line.* Sound simple? The basic concept is, but not all feedlines are the same. We've already mentioned a common type of feedline called coaxial cable. You might be familiar with another called *twinlead.* Twinlead consists of two conductors in parallel to each other, separated by a flexible insulating material such as plastic. Twinlead is a flat, ribbon-like cable; it's the cable often used with indoor TV and FM antennas such as the old "rabbit ears" models. Figure 7-3 shows a cross-sectional view of coaxial cable in part A and of twinlead in part B. Coax consists of a center conductor surrounded by an insulating layer of a material such as polyethylene. This in turn is surrounded by a braided *shield* made of a conductor such as copper. The outer layer of coaxial cable is rubber, plastic, or other insulating material that can repel water. The impedance of coaxial cable is 50Ω. Twinlead is simpler, being—as we just mentioned—two parallel conductors surrounded by plastic or other insulating material. Twinlead comes in 75Ω and 300Ω impedances. Since most transmitters come equipped only with output connectors for coaxial cable, antenna tuners equipped with twinlead output connectors are normally used with twinlead feedlines.

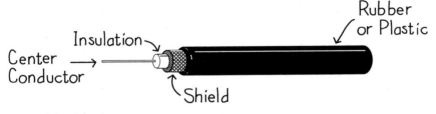

Center Conductor →

Insulation

Shield

Rubber or Plastic

A: Cross Section of Coaxial Cable

FIGURE 7-3:
Coaxial cable is more popular than twinlead, but both do the same job— transferring power from your transmitter to your antenna.

B: Twinlead

Another type of feedline you'll be quizzed about on your written exam is shown in Figure 7-4. This is called parallel conductor *open wire* or *ladder line* feedline. It is two uninsulated conductors separated by a series of insulating spacers, forming a "ladder." This feedline isn't available commercially; you'll have to build it yourself! The impedance depends on the spacing between the conductors; for example, it is 450Ω for one inch spacing and 300Ω for half inch spacing. This type of feedline must be used with an antenna tuner.

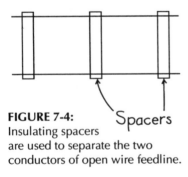

FIGURE 7-4:
Insulating spacers are used to separate the two conductors of open wire feedline.

Spacers

Coaxial cable is the most widely used kind of feedline. For one thing, all ham transmitters have a coaxial cable output, and almost all can only be used with coax unless an antenna tuner is used. Coax's impedance of 50Ω is a good match for dipoles, verticals, and beams. Coaxial cable can also be buried underground between your station and antenna.

Since the shield of a coaxial cable is kept at ground potential during operation, it is possible to run the cable along metal objects without affecting its operation. If you're like most hams, coaxial cable is the only type of feedline you'll ever use.

Parallel conductor feedlines such as twinlead or ladder line cannot be run near metallic objects, and you'll need some sort of antenna tuner to use it with your transmitter. However, they do have their advantages over coax. For one thing, less power is lost in a parallel conductor line compared to a coaxial cable of the same length. This can be important if the feed point of your antenna is more than a couple of hundred feet or so away from your transmitter. Parallel conductor lines are also more tolerant of a mismatch in the impedances of the transmitter and antenna. While such a condition should never be tolerated for long, a parallel conductor feedline can withstand it longer before being damaged than can coax.

Regardless of the type you use, some energy from your transmitter will be lost in the feedline before it reaches your antenna. This loss increases with the length of the feedline, so you should never use a longer feedline than absolutely necessary. The loss also increases with operating frequency. The same feedline will have a greater loss on 10 meters than on 80 meters. You may already know that there are different types of coaxial cables, such as "RG-58" and "RG-8" coax. These different coaxial cables have different amounts of loss for the same length. As a general rule, the larger the diameter of a coaxial cable the lower its loss. For example, RG-8 coax has a bigger diameter than RG-58, and RG-8 also has a lower loss than RG-58. The larger diameter coaxial cables are also able to handle larger transmitter powers than smaller diameter ones. A special kind of coax is RG-174, which has a smaller diameter (and lower power handling limits) than either RG-58 or RG-8. It's often used for situations where you might not want to draw too much attention to the coaxial cable, as in apartments or condominiums where outside antennas face restrictions.

Your choice of which coax to use depends on your situation. If you're operating on the HF bands from home, RG-8 is a good choice. This is especially true if you're running high power levels or your feedline is long. If you're running lower power (200 watts or less) on the HF bands, RG-58 gives satisfactory results if the feedline is short. RG-58 is a good choice for mobile installations, since it is more flexible than RG-8 and its higher losses are insignificant for the feedline length involved in a car or boat installation. RG-8 can also be used on 50 MHz. However, RG-8's losses increase rapidly at 144 MHz and higher frequencies. RG-8 can be used for short (say 25 feet or less) feedlines at VHF and UHF, but it's better to use a coax such as RG-213 which is specially designed for low losses at higher frequencies.

Baluns

Back in Chapter 5, we learned that some antennas, such as dipoles, are balanced. This means that neither side of the antenna is connected to ground. Other types, like verticals, have one side connected to ground and are said to be unbalanced. Feedlines also come in balanced and unbalanced types. Parallel conductors have neither conductor connected to ground, and are thus *balanced feedlines*. In coaxial cable, one conductor (the braided shield) is connected to ground, and so it is an *unbalanced feedline*. So what happens if we try to connect an unbalanced feedline (like coax) to a balanced antenna such as a dipole? The shield of the coaxial cable would radiate, and as a result the radiation pattern of the dipole can be distorted. To correct this, a device known as a *balun* can be added to the feedpoint of the antenna.

The term "balun" is derived from *balanced to unbalanced*. A balun will let you use an unbalanced feedline with a balanced antenna. In addition, a balun can match impedances, such as using a 300Ω parallel conductor feedline (which is a balanced feedline) with a dipole (a balanced antenna) having an input impedance of about 35Ω. If a balun is to be used to match an unbalanced feedline to a balanced antenna, the balun will be inserted at the feedpoint of the antenna. (Center connectors for dipole antennas with a built-in balun are available.) If a balun is used mainly for impedance matching, the balun is located between the transmitter and the feedline. Several antenna tuners have built-in baluns for just this purpose. Also, many beam antennas include a balun as part of their design.

To be perfectly honest, quite a few (and probably most) hams who use dipoles don't use baluns and directly connect the coax to both halves of the dipole without any apparent ill effects. While a balun should theoretically give improved performance, in actual practice the difference is often too small to be noticed. A balun can't hurt and should be used whenever it is called for, but I have to confess I've used coax to feed a dipole without benefit of a balun and have gotten away with it.

Standing Wave Ratio (SWR)

So what happens if there's an impedance mismatch between the transmitter's output and the antenna/feedline? Some of the power from the transmitter does reach the antenna and is radiated. However, a significant portion of the transmitter energy is "refused" by the antenna and is sent back down the feedline to the transmitter. This is a serious problem, since the reflected energy can badly damage the transmitter. We measure the amount of power that is actually radiated by the antenna compared to the amount reflected back by the *standing wave ratio* (SWR). The term "standing wave" refers to the fact that the energy reflected back from the antenna interacts with the energy from the transmitter to set up waveforms that seem to be "standing still" along the feedline.

The energy moving from your transmitter to the antenna is called *forward power*. The power that's reflected back down the feedline due to impedance mismatch is called *reflected power*. SWR is the ratio of forward power to reflected power as measured on the feedline. SWR can also be computed as the ratio of the maximum to the minimum voltage on the feedline or maximum to minimum currents on the feedline. The easiest way to compute SWR, however, is as the ratio between input and output impedances.

For example, let's suppose that your transmitter's output impedance is 50 Ω and the antenna impedance is 500 Ω. The standing wave ratio is 10:1. Is this an acceptable SWR? Not on your life, Bunky—it's horrendous! Since all contemporary ham radio transmitters use transistors to supply the output power to the antenna, a SWR higher than 2:1 can easily cause damage to the transmitter (in fact, some deluxe transceivers have circuits that sense when the SWR exceeds 2:1 and shut off the transmitter when it happens). A good target to shoot for is a SWR of 1.5:1 or less. A perfect impedance match between your transmitter and antenna is 1:1, but this is impossible to obtain in the real world; there will always be a little mismatch and some power will be reflected back from the antenna. Your goal should always be to keep the reflected power as low as possible. That's why many hams (including me) who use antennas such as dipoles and verticals also use antenna tuners. Since the SWR will change as transmitting frequency is changed (because an antenna is perfectly resonant—a 1:1 SWR—at only one frequency), using the antenna tuner lets you always keep the SWR as close to 1:1 as possible.

SWR is measured using a device called a *reflectometer* (although it's more commonly and informally called a *SWR meter* or *SWR bridge*). As you might suspect, the reflectometer measures the amount of power reflected back from the antenna. Most reflectometers/SWR meters are equipped with two coaxial input connectors so they can be easily inserted somewhere along the path between the transmitter and antenna. There's also usually a switch to select between forward or reflected power measurement and a meter to indicate the power or SWR. For the most accurate measurements, measurements with the reflectometer should be taken at the point where the feedline is connected to the antenna. Figure 7-5 shows an ideal arrangement for measuring SWR when an antenna tuner is also being used. But, as you might expect, this isn't always easy or practical. (Would you like to climb up a 60 foot high tower in the middle of winter—with a high wind, of course—to take a SWR measurement for a Yagi antenna???) For this reason, most SWR readings are taken with the reflectometer inserted between the transmitter output and the antenna feedline. Many antenna tuners incorporate some form of reflectometer or SWR meter. And newer transceivers (like the one AA6FW uses) have a SWR meter built in. The SWR readings you take "at the transmitter" are not as accurate as those made at the antenna, but they are better than nothing.

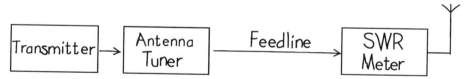

FIGURE 7-5: For the most accurate SWR measurements, the SWR meter should be placed at the antenna. So much for theory, though; this isn't a very practical arrangement!

If the SWR reading for your antenna system is 4:1 or higher, you have some real problems. Very high SWR is often a sign that there has been a component failure somewhere between your transmitter's output and the antenna, such as shorted coaxial connector or defective center insulator for a dipole antenna. If you have such high SWR, don't try to correct it with an antenna tuner! Instead, shut off your transmitter and carefully inspect the physical components of your feedline and antenna. Replace any connectors or hardware that looks suspect. When you measure SWR again after the inspection, lower your transmitter power as much as possible before transmitting again.

As we hinted a few pages ago, parallel conductor feedline is much more tolerant of high SWR than coaxial cable. The reflected power due to high SWR is dissipated as heat, and a feedline that feels warm to the touch is a strong indication of high SWR. If a high SWR condition continues, the coax conductors can get red-hot, melting the insulation and starting a fire. Fortunately (if there's anything "fortunate" about such a situation), most modern transmitters will be destroyed before coaxial cable can get that hot. Parallel conductor feedlines dissipate heat better, and are less likely to cause a fire. Of course, it's better to avoid such high SWR situations altogether!

Dummy Loads

One great way to make yourself really unpopular with your fellow hams (and maybe get a violation notice from the FCC in the bargain) is to test your transmitter with your antenna connected. Instead, you should always test your transmitter by feeding its output to a dummy load. Instead of radiating your signal, a dummy load is a jumbo resistor that dissipates your transmitter's power as heat. Dummy loads are rated according to their ability to handle transmitter power and dissipate it as heat. For example, if the maximum power from your transmitter is 200 watts, then you would need a dummy load rated at least at that power level. It's usually best to use a dummy load rated a bit more than the maximum power you expect to be transmitted. By the way, there is still some radiation from a dummy load, although it normally is extremely weak (on the order of a few milliwatts for a 100 watt input, for example). However, if the frequency is clear and another station has good receiving capabilities, you might be shocked at how far away you can be heard via the "dummy load mode." (For example, QSOs over 3000 miles have happened on bands such as 10 meters during good conditions!)

Tuning Up the Right Way

We're mentioning dummy loads in this chapter because you might be asked a question about them on your written exam and because hams are urged to "use a dummy load for tuning up instead of doing it on the air." Unfortunately, that bit of advice isn't as relevant as it used to be, nor is a dummy load as necessary as it once was. In fact, if you don't use an external power amplifier to boost your transmitter's signal, odds are you can get along fine without one.

Until about 15 years ago, virtually all ham transceivers and transmitters used vacuum tubes. When you changed bands or moved extensively in frequency (as in from one end of 80 meters to the other), you had to adjust the currents applied to the vacuum tubes that amplified your signal before it was transmitted. This was called *tuning up,* and was often an involved process that took a few minutes. While the transmitter was being tuned, it was putting out energy that had to go somewhere. Rather than put it out over the air, a dummy load could be used.

Today, all new ham transceivers use transistors to amplify signals before transmitting them. These circuits don't have to be "re-tuned" when switching bands or changing frequencies; you just connect a properly resonant antenna and away you go. Many of the high power (1500 watts output) linear amplifiers still require tuning up, so they need dummy loads (heavy duty ones at those power levels). And you should always use a dummy load when testing any transmitter. But for everyday operation? If you're using a modern transceiver, odds are you can get by just fine without a dummy load.

But today's transceivers often require a new type of "tuning up." One advantage the older vacuum tube transmitters had was that they often had the ability to operate into a wide range of antenna impedances, such as 30 to 600Ω. Today's transceivers need as close to a 50Ω antenna impedance as possible. Tuning up now means adjusting your antenna tuner so that the antenna system impedance is 50Ω. But how to do so without causing interference? Using a dummy load isn't the answer, since you're tuning your antenna system, not a dummy load.

Here's a little trick that's served me well on the HF bands. An antenna that's properly matched for transmitting is also properly matched for receiving. I tune around near the frequency I want to operate on until I find a signal that's extremely weak or an empty frequency with just background noise. I then adjust my antenna tuner until the weak signal is as strong as I can make it or the background noise is at its loudest. When I've done that, I've gotten as close to a 50Ω impedance on the antenna as I can. I check the SWR using my transceiver's built-in SWR meter, and usually it's 1.5:1 or less. After time, I get familiar with what the proper settings should be on my antenna tuner for different bands and frequencies, and the procedure only takes a few seconds. Try this instead of adjusting your antenna tuner with your transmitter on. Your fellow hams will appreciate it, and the FCC won't cite you for unidentified transmissions or causing deliberate interference!

Filters and Interference Reduction

In radio, a *filter* is a circuit or device that will permit certain frequencies to pass but reject others. Your receiver (or the receiver section of your transceiver) has filters which help you receive stations only on the frequency you want to. Other filters help you reduce interference to television sets and other electronic devices caused by energy from your transmitter. Due to the poor design of many consumer electronic products, it's possible that you could cause interference to your neighbor's TV or stereo system even if you're doing nothing wrong.

There are three basic types of filters you should be familiar with. Figure 7-6 has graphs of signal frequency versus strength which show how these filters affect signals passing through them. In part A of Figure 7-6, signals below a certain frequency can pass unimpeded through the filter. But when the signals reach a certain point, called the *cutoff frequency*, the filter will greatly attenuate (reduce in strength) all signals higher in frequency. This type of filter is called a *low pass filter*, since frequencies below the cutoff point pass without attenuation. Now look at part B of Figure 7-6. This looks like the graph in part A turned on its head—signals below the cutoff point are attenuated, while frequencies above it pass easily. This kind of filter is known as—you guessed it!—a *high pass filter*. Finally, part C shows a filter that blocks signals above and below a certain range of frequencies while allowing that certain range to pass without attenuation. This type of filter is called a *bandpass filter*, and the range of frequencies it passes is called a bandpass.

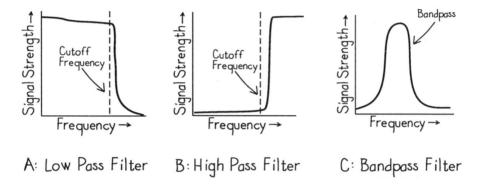

A: Low Pass Filter   B: High Pass Filter   C: Bandpass Filter

FIGURE 7-6: Here are the "response curves" for the three most common filter types. The bandpass filter is why your receiver can tune in just one station at a time instead of every station on the band!

Back in Chapter 3, we took a look at harmonics. As you remember (sure you do!), harmonics are integer multiples of a frequency. For transmitters operating on frequencies below 30 MHz, this means that harmonics from them can show up in the TV and FM broadcast bands, causing interference to reception. To prevent such harmonics from being radiated, a low pass filter can be installed between the transmitter's output and the feedline. The usual cutoff frequency is about 30 MHz, meaning all frequencies below that can pass without trouble but all higher frequencies are suppressed. Low pass filters are available as separate accessories, and most currently available ham transceivers come equipped with a built-in low pass filter.

However, your transmitter can be radiating no significant harmonics and yet still be causing interference to a neighbor's television set or stereo receiver. This can happen when the TV or stereo receiver is connected to an ordinary antenna (not to a cable system) and isn't selective enough to reject the energy from your transmitter; your radio signals can "overpower" your neighbor's devices and cause interference. If this happens, you can often alleviate the problem by installing a high pass filter with a cutoff frequency of 30 MHz or so between the TV or stereo receiver and the antenna. Such a filter will block energy from your transmissions below 30 MHz and keep them from causing interference. However, should your neighbor be connected to a cable television system, *don't* add a high pass filter or other device to the set. Let the cable company handle the problem, since they are responsible for keeping their cable system "RF tight."

How can you tell if the problem lies in your transmitter or your neighbor's stuff? One good sign is whether the interference is present across the FM broadcast band and on all TV channels or if it's restricted to just one or two channels and spots on the FM band. In the former situation, your signals are overloading their equipment and the problem lies in the design of their items. A high pass filter can help in such cases. If the latter situation is happening, however, then your transmitter is radiating harmonics and you need to install a low pass filter at your transmitter. If other consumer electronics products—such as telephones, CD players, or VCRs—the problem is overloading.

Interference caused to consumer electronic devices due to ham radio operation is known by the generic terms *radio frequency interference* (RFI) or *television interference* (TVI), even if the item suffering interference is a telephone or CD player. This is a problem of growing seriousness, as many consumer electronics items are not sufficiently selective to reject interference from ham transmitters legitimately operating within FCC regulations. As you might expect, though, few people want to be told that it's *their* TV set or stereo that's at fault—it has to be that *#@!! down the street with his or her ham radio set that's to blame!

Resolving TVI cases, even if you're not at fault, is an important task for any ham. The key tool to keeping good relations with your neighbors is your sense of cooperation and goodwill. Don't ignore TVI complaints; instead, take them seriously and work with your neighbor(s) to reach a solution everyone can live with. Solving TVI problems is beyond the scope of this work, but the ARRL (and others) have some informative books on the subject. Local ham clubs often have "TVI committees" that work with the public on cases of TVI. You can help prevent TVI problems by using no more transmitter power than necessary (there are some hams who use the full 1500 watts of power even if they're just communicating with the next state!), connecting a heavy gauge wire from your transmitter's ground connection terminal to a stake driven into the Earth, and making sure that your transmitter is not radiating harmonics or overmodulating. FM tends to cause less TVI than AM and SSB, and VHF and UHF bands such as 2 meters are less likely to cause TVI than operation on HF bands.

Bandpass filters for receivers or the receiving sections of transceivers are rated in kHz, which indicates how broad a range of frequencies can pass through the filter before being rejected. For AM, a bandwidth of 4 to 6 kHz is good. SSB signals take up less frequency space than AM, so bandpass filters rated at 2.3 to 2.9 kHz give good performance. These narrower filters can often be used for AM reception; however, the audio will sound "clipped" and have less fidelity since some of the two sidebands of an AM signal are being rejected. CW filters can be very narrow—500 Hz is a common width. For FM, a filter will need to be at least as wide as the deviation above and below the unmodulated frequency. Since ham FM has a deviation of 5 kHz, FM bandpass filters will need to be at least 10 kHz wide. The filters in most ham FM transceivers are rated at 15 kHz. This accommodates the necessary deviation range plus a "reserve" for signals that are overmodulating a bit.

Test Equipment and Measurements

There are several items of test equipment which hams find useful. Some of these (such as *RF wattmeters* to indicate transmitter output power) are available as separate units or as built-in features of transceivers and antenna tuners. Other items of test equipment, such as those to measure voltage and current, are always separate items of gear. While you don't need an extensive set of test and measurement gear to get on the air initially, you'll probably wind up collecting quite a few pieces over the years! But, for now, here's what you need to know for your written exams.

As just mentioned, RF wattmeters measure the output power of transmitters. The output power of a transmitter is measured by connecting a RF wattmeter directly to the output connector of the transmitter. Most RF wattmeters are designed with a 50Ω impedance, so there's usually no connection problem. Most RF wattmeters are of the *peak-reading* type, meaning they measure transmitter power at the peak (or crest) of the modulation "envelope." Figure 7-7 shows a sine wave signal with the peak of the modulation envelope indicated. Power measured at this point is referred to as the *peak envelope power* (PEP) of the transmitter. Novices are permitted to use up to 200 watts PEP of output power on the bands below 30 MHz, 25 watts PEP on 220 MHz, and 5 watts on the 1270 MHz band. Technicians (and all other license classes) are allowed up to 1500 watts PEP above 30 MHz.

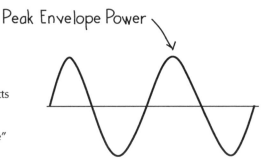

Peak Envelope Power

FIGURE 7-7: When the FCC says Technician and higher class hams can use 1500 watts PEP of power, they mean 1500 watts measured at the peak of the signal "envelope" as shown here.

Some RF wattmeters are *directional*, meaning they can measure either the power from the transmitter to the antenna or the power being returned from the antenna to the transmitter. The actual output power of a transmitter is the forward power from the transmitter less any energy reflected back from the antenna. Suppose that you measure 100 watts in a forward direction and 5 watts reflected back from the antenna. The actual transmitter output power in this case is 95 watts.

Many contemporary transceivers come equipped with built-in meters to measure both RF power output and SWR. Many antenna tuners also include wattmeters. Some RF wattmeters even include two indicating needles—one to indicate power in a forward direction, the other for reflected power, with both operating simultaneously.

Often hams need some way of checking the calibration accuracy of the frequency indicator dial of a transceiver or receiver. A *marker generator* is used for this purpose. A marker generator will produce a tone at precise fixed frequency intervals, such as 100, 25, or 5 kHz, throughout the amateur bands. Many transceivers and receivers let you slightly adjust the calibration of the frequency indicator dial and "zero" it on a marker generator tone. A marker generator is also handy in tuning to a desired frequency if you have trouble reading the frequency indicator dial (such as when operating from your car and you can't take your eyes off the road). In some older transceivers and receivers, this circuit was sometimes referred to as a *crystal calibrator*.

Another way to test the accuracy of a transceiver or receiver frequency indicator dial is with an *RF signal generator*. This device produces a weak radio signal on a wide range of frequencies (one I own can produce a signal on any frequency between 100 kHz to 150 MHz, for example). Some RF signal generators come equipped with a highly accurate digital frequency display; others can be used in conjunction with a device known as a *frequency counter* which has a digital frequency display accurate to 1 Hz. (I use a separate frequency counter with my RF signal generator.) A RF signal generator and a frequency counter are really useful items. I use mine for checking the calibration accuracy of my transmitting and receiving equipment, performing maintenance (called *alignment*) on some of my older vacuum tube equipment, and for tests and experiments where I need highly accurate signals of a known frequency. You won't need these items right away—in fact, many hams don't own either—but I now wonder how I got along without them for so many years.

The one item you will need is called a *multitester* or *multimeter*. In electronics, this device is the equivalent of a Swiss Army knife. Multitesters can measure voltage, current, resistance, as well as check for circuit continuity (whether a circuit is open or closed). Multimeters come in a wide range of prices, with better units able to measure larger voltages and currents. They also come with analog dials (a needle or pointer against a scale) or with digital readouts. The digital readout models are easier to use, but the analog dials are often better at reacting to quick changes in the circuit you're measuring, such as rapidly fluctuating voltages. Either type will do fine for your first multitester, but eventually you'll probably want to have one of each (like I do). Figure 7-8 shows a typical multimeter available from Radio Shack. Virtually all ham radio equipment dealers and electronics supply companies also have several models of multitesters for sale.

FIGURE 7-8: A multimeter, such as this one available from Radio Shack, should be the very first item of test equipment you buy.

All resistance measurements and continuity checks are done with the power to a circuit turned off. You just place the two probes at the points in the circuit where you want to measure the resistance across or determine whether the circuit path is open or closed. To measure voltage or current, power must be applied to the circuit. Voltage is measured by placing the probes so that they are in *parallel* with the circuit or device where voltage is being measured. To measure current, do the opposite—place the probes so they are in series with the circuit or device where you're measuring current.

It's possible to extend the range over which your multi-tester can measure voltage or current. To extend the voltage measuring range, you would add a resistance in *series* with the multitester. Using Ohm's law, you can see that resistances of different values will deliver smaller voltages to the multitester, allowing it to "measure" voltages beyond its normal range. (However, the dial scale would no longer be accurate, as it would indicate a lower voltage level than is actually being measured.) To extend the current measuring range, you'd do the reverse—you would add resistance in *parallel*. The resistance will deliver a lower current to the multitester than is actually flowing in the circuit, allowing a multitester with a lower maximum current measuring capability to be used. In actual practice, few hams (maybe none???) ever add a resistance in series or parallel with their multitesters to extend their measuring range. However, multitesters use this same basic principle to allow you to measure different ranges of voltages and currents. A basic meter circuit to measure a relatively low range of voltage or current is built into the multitester. A rotary switch lets you select higher ranges of voltages, currents, and resistances; each turn of the switch selects a different resistance appropriate for the selected range to be added in series or parallel with the meter circuit. Some deluxe multimeters include circuits which automatically switch in the correct series or parallel resistance for the voltage or current level being measured. These are called *autoranging* multimeters.

Trying to get the most out of ham radio without a multi-tester is like trying to read in the dark. Make a multitester the very first item of test and measurement gear you add to your shack!

Strange Technical Lingo

Ham radio shares many words with other areas of electronics. But it also has a set of technical terms and slang that's unique to it. Even if you've had some prior experience with electronics, some of these might be new to you. . . .

active filter: a circuit that takes the audio output of a receiver and eliminates audio frequencies not wanted. This is especially useful in CW reception (since the output of a CW receiver is just one audio tone) and with SSB reception, where audio frequencies higher than 2700 Hz or so aren't needed for intelligible speech.

AGC: short for "automatic gain control," a circuit that adjusts the amount of radio frequency and audio amplification given a received signal so that the output from the receiver's speaker stays at a fairly constant volume level.

base-loaded: when the loading coil is located at the bottom of a vertical antenna.

birdie: a false signal in a receiver produced as a by-product of the receiver circuitry.

center-loaded: when the loading coil is located at the center of a vertical antenna.

circular polarization: an antenna design where polarization "rotates" between vertical and horizontal.

crystal-controlled: when the frequency of an oscillator or transmitter is determined by a quartz crystal resonant at the desired frequency. Such circuits offer good frequency stability.

drift: when a transmitter or oscillator unintentionally varies off its proper operating frequency.

ERP: short for "effective radiated power," which is the transmitter power and antenna gain considered together.

frequency synthesis: a method of tuning a transmitter or receiver by combining several crystals to produce a wide range of frequencies.

general coverage: a receiver (or receiving section of a transceiver) that can cover a wide frequency range, such as 100 kHz to 30 MHz, and operate in different modes such as AM, SSB, FM, CW, and RTTY.

hollow-state: a spoofing term for a vacuum tube circuit or device. Compared to solid-state circuits, vacuum tubes use more power, are bulkier, give off more heat, and tubes fail more often. However, vacuum tubes are still often the most cost-effective way to amplify signals at high power levels.

integrated circuit: a complete circuit in miniature etched onto a small chip of silicon.

load down: when a device (such as an item of test equipment) draws too much current from a circuit, causing the voltage to drop and the measurement to be inaccurate and circuit performance to suffer.

noise blanker: a receiver circuit designed to minimize the effects of certain types of electrical noise.

notch filter: a circuit that takes a small slice, or "notch," out of the bandpass of a receiver. This is handy for removing a narrow bandwidth interfering signal, such as a CW signal when you're trying to hear a SSB station.

parasitics: in an oscillator, oscillations on a frequency other than the desired one.

passband tuning: a circuit in a receiver that allows adjusting the bandpass of the receiver for best performance under various conditions of interference.

PLL: short for "phase locked loop," a circuit that can generate a wide range of different frequencies in discrete steps such as 10 or 100 Hz. This is the most widely used technique of controlling transmitting and receiving frequencies in today's ham radio equipment.

preamp: a receiving circuit that gives extra amplification to very weak signals; however, it also increases the background noise.

rockbound: same as crystal-controlled, but usually spoken with a negative tone. "I can't go to another frequency. I'm rockbound."

rubber ducky: a inductively loaded, flexible antenna for use at VHF and UHF.

S-meter: a meter or bargraph (i.e., a row of lights) used to indicate the relative strength of received signals.

skyhook: an antenna—especially a large, grotesque one that upsets the neighbors and frightens small children.

smoke test: the ultimate reality check as to whether or not a circuit works or a repair was made correctly. Apply power to the circuit or device. Does it smoke? If it doesn't, you did okay. If it does, you didn't.

solid-state: a circuit that uses no vacuum tubes, only transistors, integrated circuits, and similar tiny devices made from silicon. These circuits are smaller, lighter, use less power, and are less prone to failure than vacuum tube circuits.

speech processing: a circuit that boosts the average level of the modulating signal applied to a transmitter so that the average percentage of modulation in the transmitted signal is higher. However, it distorts the audio a bit, and if overdone can make it impossible to understand the signal on the receiving end.

spurs: unwanted signals in the output of a transmitter or oscillator, often distorted and on frequencies other than the intended one. Can cause interference.

top-loaded: when the loading coil is located at the top of a vertical antenna.

translator: a device that receives an entire range of frequencies (such as 100 kHz of the 2-meter band) and relays the entire range on another band (such as 10 meters.)

transverter: a circuit that takes the output of a transmitter on a certain band, such as 10 meters, and converts it to another band, such as 2 meters. On receive, the process is reversed; signals received on a band such as 2 meters are converted to another, such as 10 meters, and fed to a receiver tuned to 10 meters. A transverter converts only one signal at a time; a translator converts an entire range at once, and that range could contain several signals at the same time.

VFO: short for variable frequency oscillator, the circuit used to set the receiving or transmitting frequencies of ham gear.

VOX: a circuit that turns on the transmitter automatically whenever you speak into the microphone without having to press any button or switch. These circuits were originally developed for use in SSB operation, but most hams just press the microphone button when they want to talk and forget about VOX.

Safety

Contrary to an unfortunate perception by some members of the public, hams don't have electricity zapping around their radio rooms like a scene from the *Frankenstein* films of the 1930s. Fortunately for us, ham radio is one of the safer "active" (as opposed to, say, coin or baseball card collecting) hobbies. However, it is possible to cause yourself and others serious injury (or even death) if you ignore common sense and some basic safety procedures. Let's take a look at some of the possible dangers and how to protect yourself.

It doesn't take a great deal of voltage or current to cause death. The key is whether or not the electricity has a path through the heart tissue. If it does, as little as 30 volts or 100 milliamperes can be fatal; 50 milliamperes can be quite painful. As a general rule, you don't have much to worry about when using or working on battery-powered equipment (even though a 12-volt battery can deliver a jolt you'll remember a long time). However, you must exercise extreme caution whenever working with a higher powered transmitter (about 50 watts or more), even if it is powered from a battery, since high voltages can be developed inside such a transmitter. Of course, *any* device that's powered by the AC line must be treated with extreme care. All power to such devices must be shut off before you open their cabinets or attempt to work on them. Some transmitters and power supply units are designed with a *safety interlock* built in. The safety interlock will shut off power to the circuit immediately if you try to open the cabinet with the power switched on. (However, the result is a loud popping noise as the interlock is tripped, and opening an equipment cabinet with the AC power applied isn't recommended.) *Don't* try to defeat the safety interlock by using a "cheater" power cord or other method to keep the circuit powered with the cabinet removed. Unfortunately, every year at least a couple of hams get electrocuted trying to get around the protection of a safety interlock. And even if you have the cabinet removed and the power off, dangers still lurk. For example, high voltage capacitors can remain charged long after the power is removed and be capable of delivering a lethal shock. This is especially true of capacitors used in power supplies. The golden rule when working with AC line powered equipment is simple: *don't, if you don't know what you're doing!* It's great fun to learn to work on your own equipment if you have a more experienced ham as a mentor, or as part of a course at a local community college or vocational education center. But don't go poking around inside ham equipment if you don't know what you're doing. You'll probably do more harm than good, and you might fry yourself in the process. (Have I scared the daylights out of you? Good—that was my intent.)

It's important that you always know where the main power switch of each piece of your equipment is and that your equipment be placed so you can easily reach the switch so that you can quickly shut it off in case of a problem. It's also good for you and all members of your family to know where the main power switch is for your house or the room where you have your ham gear. If someone should be unconscious and in contact with some of your ham gear, it's best to first turn off the power before attempting to move that person—would-be rescuers could be electrocuted if the power is still on. And if you have kids or others in your home who might go poking around your ham station when you're not around, it's a good idea to install some sort of key-operated on/off switch in the power line to your radio room or equipment. (Let me make a little confession here: the written exams are fond of asking a question about key-operated power switches for prevention of unauthorized operation, so remember this little tidbit. But I've yet to run across any ham who actually uses one!)

If you ever do some work on a circuit that connects to an AC line (such as replacing a power cord), you need to know the color coding used on three-wire (i.e., three prongs on the wall socket) AC systems. The green wire is the ground connection wire, and this normally goes to the chassis of the circuit or device. The black (sometimes red) wire is the "hot" wire. This is the wire that has the on/off switch and any fuse used in the circuit. As you might expect from its name, this is one wire you shouldn't touch! The remaining wire is white and is called the neutral wire. You shouldn't touch this wire either. Color coding even shows up in places like the retaining screw of a light socket; the silver thread of the screw is neutral and the brass "button" is the hot connection.

Radio energy can also cause injury since it is a high frequency form of AC. You should always locate antennas so that no one can accidentally come into contact with it. (The same also applies if you're using "ladder line" feedline.) Moreover, RF energy, especially at VHF and UHF frequencies, can cause injury even if you don't come into direct physical contact with an antenna. Do you have a microwave oven in your home? It works by sending UHF radio energy through your food, causing it to heat. Energy from radio transmitters operating above 300 MHz can do the same thing to the human body if you're careless. The eyes are very susceptible to injury in this way; I've known hams who have used hand-held UHF units for lengthy periods who have experienced dry, "bloodshot" eyes the next day! The American National Standards Institute (ANSI) recommends that the power output of hand-held VHF/UHF transmitters be limited to 7 watts or less, and as a result ham hand-held transceivers limit their power output to 5 watts or less. When using a VHF/UHF hand-held unit, always hold it so that the antenna is pointing away from your head when you're transmitting. If you install a VHF/UHF transceiver in your car, you should position the antenna away from the driver's head and those of your passengers. The middle of the roof and the trunk are two good locations.

It's also a good idea to install VHF/UHF antennas at your home station in a location where no one can get near them when you're transmitting. Use high quality connectors and coaxial feedline at VHF and UHF frequencies, as these help prevent leakage of RF energy.

Speaking of antennas for any frequency, probably more hams injure themselves when their antennas are *not* being used for transmitting than when they are! This can most often happen when climbing a tower supporting a beam antenna. To prevent falls and the injuries they can cause, a safety belt should be worn by anyone who climbs a tower. People on the ground helping out can be injured if a tool or piece of the antenna falls, so they should wear a hard hat and safety glasses.

Perhaps the biggest danger facing any ham with an outside antenna involves lightning. Obviously, a direct lightning strike hitting your antenna can do tremendous damage to the antenna, any equipment connected to the antenna, and your house if the lightning strike follows the feedline from your antenna into your house. Even a nearby strike that doesn't hit your antenna can induce (remember our discussion of inductors and the relationship between electricity and magnetism?) strong currents in your antenna capable of damaging your equipment. One obvious means of protecting your station (and home) is to disconnect all of your station equipment from AC lines and antennas when not in use. But for maximum protection you need to "ground" all of your station equipment and antenna structures.

We've looked at a couple of meanings of the term ground earlier in this book. Here we are talking about a direct physical connection to the Earth itself. The connection is made to an 8 foot or longer copper or copper-clad rod driven into the ground. Heavy gauge braided flat copper wire known as "grounding strap" should be used for all connections (the shield from larger diameter coax such as RG-8 can also be used). *All* station equipment should be connected to or be capable of being connected to this ground rod. For example, you could use a SPDT switch to connect your feedline and antenna to the ground rod when not in use. There are also special devices known as *lightning arrestors* which can be installed in the feed-line for an antenna. These are connected directly to the ground rod and send any voltages over a certain level to ground. In your station, connect all items of equipment—transceivers, antenna tuners, linear amplifiers, etc.—to ground. You can make connections to a common point in your station (usually your transceiver or transmitter) and then run the grounding strap from that point to the ground stake. Besides lightning protection, a good ground connection can help prevent RFI/TVI problems. Ground connections are also important on the lower frequency bands such as 80 and 40 meters, and supposedly help to reduce noise when receiving—although I've never known the latter to be true!

By the way, you might have a question on your written exam about indoor grounding points for your station. The correct answer for such a question will be "an indoor cold water pipe." For years, cold water pipes in homes were made of copper and provided a direct path from the house to ground. That may still be true for older homes, but it's seldom so with newer houses. Much of the plumbing for new construction makes heavy use of polyvinyl chloride (PVC) pipes and joints; this is true even if the plumbing you can see is copper. Some contractors use copper for the visible plumbing because people associate it with quality, but the plumbing behind the walls is PVC. Regardless of what the "correct" answer is on your written exam, I suggest you forget about using a cold water pipe as a ground point unless you *know* that a complete electrical path exists from the connection point to the Earth. And don't make a connection to a natural gas pipe or a hot water pipe. In the latter case, there's a shock hazard from the water heater.

I hope I haven't frightened you away from ham radio with this section on hazards. The vast majority of hams go their entire "ham career" without any incident of any kind (not even an accidental shock strong enough to be noticed). But some hams do die each year as a result of electrocution, falling off a tower, etc. What makes these deaths especially horrible is that in almost every case they were preventable if the practices we've just discussed had been followed.

If I do nothing else in this book, I hope I've conveyed the importance of safety to you. I want you around for a lot of years so you can enjoy ham radio—and maybe buy some more of my books!

CHAPTER EIGHT

Hey! We're Getting Near the Finish Line!

I F YOU'VE MADE IT THIS FAR in this book, you're 90% done in your quest for your ham license. Take a minute and congratulate yourself!

In this final chapter, we're going to take care of some remaining business necessary to get you over the top and on the air, and then look at selecting your first station, getting on the air for the first time, and what to do once you're up and running. We'll be returning to some of the same territory we covered back in chapter one, so it might be a good idea to re-read it before beginning this one.

Get Your Ham License Someday

Yeah, get it someday. And today is that someday!

If you've read this far in this book, you're obviously interested in ham radio. If you've understood most (it doesn't have to be everything) of what you've read so far, you're capable of passing both the Novice and Technician written exams.

Make up your mind now that you're going for the Technician class license. Forget the Novice—its privileges are just too limited and frustrating. Those 25 additional questions that make up the Technician written exam give you an incredible boost in your operating privileges and capabilities. If you can learn the Morse code at the 5 WPM required for Novice privileges, great! Take the code test at the same time you take the Technician written exam and you'll have Novice HF privileges added on to your Technician license. But go for that Technician license right out of the gate.

Don't wait until you've invented enough excuses why you can't pass the exams. Today—before you have a chance to talk yourself out of it—drop a note to the Federal Communications Commission, Gettysburg, PA, 17326 and ask for a copy of form 610, the application form for an amateur radio license. Figure 8-1 shows what it will look like. While you have your pen, envelopes, and stamps handy, drop a note (along with a stamped, self-addressed envelope) to both the ARRL and W5YI volunteer exam coordinators (VECs), asking who gives exams in your area. You might also want to drop another note to the ARRL asking for the addresses and contact people for ham radio clubs in your area and membership information for the League. When you get a name or two from the ARRL for clubs in your area, call the person indicated and tell them you're interested in getting your ham radio license and would like to attend the next meeting. Then go to it. Many local radio clubs offer ARRL membership as part of their annual dues. If they do, join ARRL through them. If not, join a local club and then join ARRL separately. While you're waiting for the ARRL to process your membership application, drop by your public (or college) library and read the latest issue of QST. If the library doesn't have it, ask the staff about getting a subscription to it. Check the library for any books they might have on ham radio (if they don't have this one, tell them to get it!) and see if you can find copies of 73 and CQ at larger newsstands.

When you get the information from the ARRL and W5YI VECs, and you have a copy of form 610 from the FCC, make a date about a month later to take an exam. That's your target, and it can focus your mind wonderfully on the task at hand. You'll need additional study materials (such as the study guides mentioned back in Chapter 1), and they're available from the same place you got this book. Set up a study schedule for yourself and stick to it. If you can spare an average of a half-hour a day for a month before your exam, you'll be able to pass the Novice and Technician exams. I believe you can do it. You just have to believe it too.

In short, don't kid around. If you're serious about getting a ham license, don't try to ease your way into ham radio—jump in with both feet!

FEDERAL COMMUNICATIONS COMMISSION
GETTYSBURG, PA 17326

Approved OMB
3060-0003
Expires 12/31/92
See instructions for information
regarding public burden estimate

APPLICATION FOR AMATEUR RADIO
STATION AND/OR OPERATOR LICENSE

ADMINISTERING VEs' REPORT			EXAMINATION ELEMENTS							
Applicant is credited for: ➡			1(A)	1(B)	1(C)	2	3(A)	3(B)	4(A)	4(B)
A. CIRCLE CLASS OF FCC AMATEUR LICENSE HELD: N T G A	Class ➡		(NT)	(GA)		(NTGA)	(TGA)	(GA)	(A)	
B. CERTIFICATE(S) OF SUCCESSFUL COMPLETION OF AN EXAMINATION HELD: ➡			Date Issued	Date Issued	Date Issued	Date Issued	Date Issued	Date Issued	Date Issued	Date Issued
C. FCC COMMERCIAL RADIOTELEGRAPH OPERATOR LICENSE HELD:	Number:									
	Exp. Date:									
D. EXAMINATION ELEMENTS PASSED THAT WERE ADMINISTERED AT THIS SESSION: ➡										

E. APPLICANT IS QUALIFIED FOR OPERATOR LICENSE CLASS: □ NONE:	H. Date of VEC coordinated examination session:
E1. □ NOVICE (Elements 1(A), 1(B), or 1(C) and 2)	
E2. □ TECHNICIAN (Elements 1(A), 1(B), or 1(C), 2 and 3(A))	I. VEC Receipt Date:
□ GENERAL (Elements 1(B) or 1(C), 2, 3(A), and 3(B))	
□ ADVANCED (Elements 1(B) or 1(C), 2, 3(A), 3(B) and 4(A))	
□ AMATEUR EXTRA (Elements 1(C), 2, 3(A), 3(B), 4(A), and 4(B))	

F. NAME OF VOLUNTEER-EXAMINER COORDINATOR: (VEC coordinated sessions only)

G. EXAMINATION SESSION LOCATION: (VEC coordinated sessions only)

SECTION I

1. IF YOU HOLD A VALID LICENSE ATTACH THE ORIGINAL LICENSE OR PHOTOCOPY ON BACK OF APPLICATION. IF THE VALID LICENSE OR CERTIFICATE OF SUCCESSFUL COMPLETION OF AN EXAMINATION WAS LOST OR DESTROYED, PLEASE EXPLAIN.

2. CHECK ONE OR MORE ITEMS, NORMALLY ALL LICENSES ARE ISSUED FOR A 10 YEAR TERM.

2A. □ RENEW LICENSE – NO OTHER CHANGES ➡	EXPIRATION DATE (Month, Day, Year)
2B. □ REINSTATE LICENSE EXPIRED LESS THAN 2 YEARS ➡	
2C. □ EXAMINATION FOR NEW LICENSE	
2D. □ EXAMINATION TO UPGRADE OPERATOR CLASS	FORMER LAST NAME SUFFIX (Jr., Sr., etc.)
2E. □ CHANGE CALL SIGN (Be sure you are eligible – See Inst. 2E)	
2F. □ CHANGE NAME (Give former name) ➡	
2G. □ CHANGE MAILING ADDRESS	FORMER FIRST NAME MIDDLE INITIAL
2H. □ CHANGE STATION LOCATION	

3. CALL SIGN (If you checked 2C above, skip items 3 and 4)	4. OPERATOR CLASS OF THE ATTACHED LICENSE:

5. CURRENT FIRST NAME	M.I.	LAST NAME	SUFFIX (Jr., Sr., etc.)	6. DATE OF BIRTH ___/___/___ MONTH DAY YEAR

7. CURRENT MAILING ADDRESS (Number and Street)	CITY		STATE	ZIP CODE

8. CURRENT STATION LOCATION (Do not use a P.O. Box No., RFD No., or General Delivery. See Instruction 8)		
	CITY	STATE

9. Would a Commission grant of your application be an action which may have a significant environmental effect as defined by Section 1.1307 of the Commission's Rules? See instruction 9. If you answer yes, submit the statement as required by Sections 1.1308 and 1.1311. □ YES □ NO

10. Do you have any other amateur radio application on file with the Commission that has not been acted upon? If yes, answer items 11 and 12. □ YES □ NO

11. PURPOSE OF OTHER APPLICATION	12. DATE SUBMITTED (Month, Day, Year)

CERTIFICATION

I CERTIFY THAT all statements herein and attachments herewith are true, complete, and correct to the best of my knowledge and belief and are made in good faith; that I am not a representative of a foreign government; that I waive any claim to the use of any particular frequency regardless of prior use by license or otherwise; and that the station to be licensed will be inaccessible to unauthorized persons.

WILLFUL FALSE STATEMENTS MADE ON THIS FORM OR ATTACHMENTS ARE PUNISHABLE BY FINE AND IMPRISONMENT
U.S. CODE TITLE 18, SECTION 1001

13. SIGNATURE OF APPLICANT: (Must match Item 5)	14. DATE SIGNED:

(OVER)   FCC Form 610, February 1990

FIGURE 8-1: This is a form familiar to all hams and prospective hams—form 610, the FCC application for a ham radio license.

What Should My First Station Be Like?

Good question! Take a look inside any ham radio magazine or store and your eyes will glaze over at all the different items of equipment you can buy. Buying those first pieces of equipment for your station usually involves some tough decisions and compromises, especially if you're on a tight budget.

I can't tell you what you should do. But I can tell you what I'd do if I was in your position.

Code-Free Technician

If I had just passed the Novice and Technician written exams for my code-free Tech ticket, my first move would be to buy a hand-held 2-meter FM transceiver like the one shown in Figure 8-2. You can pick up a new model with all the bells and whistles you need for under $300, and you'll have a complete station for voice or packet operation on the most popular ham band. You can use such a unit at home (you can use it with an outside 2-meter antenna), in your car (again with an external antenna), or as a portable unit. It will let you get involved in the mainstream of ham radio—participating in such activities as packet, public service, and general chit-chat—and keep you busy and happy until you get a better idea of which areas of VHF and UHF ham radio you're interested in most. If you're like a lot of hams, you'll wind up using this unit more than any other piece of equipment in your shack. Mine is my constant traveling companion, and I often leave it on to monitor local repeaters while I'm doing something else. A 2-meter FM hand-held unit is the closest thing to a membership card in ham radio—all real hams holding a Technician or higher class license have one.

My next purchase would depend on whether or not I owned a personal computer (especially a MS-DOS system) and my available funds. If I did own a PC, I'd add a TNC so I could operate packet radio. For frequencies above 30 MHz, a packet-only TNC instead of a multimode (packet, RTTY, CW, etc.) unit will do fine since virtually all activity on those frequencies is packet. But if I had any ambitions of maybe getting active on frequencies below 30 MHz, or if I owned a receiver capable of tuning the ham bands below 30 MHz, I'd get a multimode unit.

What if I didn't own a microcomputer? This is a close call, but my next purchase would be some sort of shortwave receiver capable of tuning the HF ham bands. This receiver wouldn't have to be some sort of monster costing several hundreds of dollars; a portable shortwave receiver under $300 would do fine as long as it met two requirements: 1) it has a digital dial to indicate the exact frequency that it's tuned to, and 2) it has the capability to receive SSB signals. If I did own a PC, a shortwave receiver would be my next purchase after a multimode TNC. Why add receiving capability below 30 MHz if I had a license that limited me to transmitting above 30 MHz? For

FIGURE 8-2:
If you get a Technician license, this might be all the ham radio gear you'll ever need—a 2-meter FM walkie-talkie such as this model from Radio Shack. No other piece of gear will give you as much fun for the money.

one thing, it would let me tune in stations such as WWV and W1AW and the services they provide. If I have a microcomputer and multimode TNC, I can copy plenty of RTTY traffic, including W1AW bulletins. I would also have plenty of CW QSOs at all sorts of speeds to copy in an attempt to build up my code speed. Finally, a shortwave receiver would let me listen in on the world of HF ham radio even if I couldn't fully participate in it. (Dropping the hypothetical for a moment, I have in the past lived in a high-rise steel frame condominium where I couldn't install any sort of outside antenna nor radiate any sort of effective signal on the HF bands. I was restricted to 2-meter FM operation using a mobile antenna placed near a window. If I hadn't been able to at least listen to the HF bands, I would've gone nuts!)

FIGURE 8-3: The Kenwood R-5000 is a high performance shortwave receiver. Even if you restrict your operation to the bands above 30 MHz, a receiver like this can be an important part of your station.

So those are what I would consider necessities for a new Technician class licensee—a 2-meter FM hand-held, a TNC (if I owned a PC), and a shortwave receiver with a digital dial and SSB capabilities so I could listen to the bands below 30 MHz. What next? Now the choices get a bit tougher. I would probably next add a higher powered (25 watts or so) SSB/FM/CW 2-meter transceiver for my home station along with the best antenna system I could erect (hopefully, a Yagi with considerable gain). This arrangement would be ideal for packet, FM simplex, SSB DXing via tropo and sporadic-E, and as the core of a 2-meter uplink for satellite communications.

And next? You've got me! There are a lot of different routes to go from here. I'd have a hard time choosing between a SSB/FM/CW transceiver and matching antenna for 50 MHz or the same for 420 MHz. The 6-meter rig would let me work worldwide DX (maybe even over 100 countries at the sunspot cycle peak) and provide plenty of sporadic-E openings each year. But the 70 cm transceiver would put me on the second most popular band above 30 MHz. There are repeaters, tropo DXing, television, and experimenting up there, and a 420 MHz transceiver could be paired with the 2-meter SSB/FM/CW transceiver to make a basic satellite communications system. I would probably avoid getting on the 222 or 1240 MHz ham bands until I had some capability on 420 MHz. There's just too much interesting and fun stuff to do on 70 cm compared to those two bands.

Two Good Reasons to Own a Shortwave Receiver: W1AW and WWV

Every ham, even those who operate strictly on the VHF/UHF bands, needs to be able to tune the HF bands below 30 MHz. Two good reasons are W1AW and WWV.

W1AW is the call sign of a ham radio station—the one at ARRL headquarters in Newington, CT. W1AW was the call sign of Hiram Percy Maxim, the founder of the ARRL, and was issued to the League upon his death as a memorial to him. In keeping with Maxim's spirit, W1AW continues to provide many valuable services to ham radio.

Perhaps the most important service W1AW provides is the daily bulletins of material of interest to hams. These bulletins include news about ham radio, such as the latest FCC proposals, DX stations being heard, and propagation forecasts. These bulletins are transmitted daily in CW, RTTY, ASCII, AMTOR, and SSB. W1AW also is the station where any news about a communications emergency (such as those involving a major natural disaster) is first announced. If you're interested in building up your Morse code speed, W1AW has daily code practice transmissions at precise speeds from 5 to 35 WPM. W1AW operates on all bands from 160 to 2 meters, and a complete schedule and current frequencies can be found in a recent issue of QST.

WWV is a government station located at Fort Collins, CO. WWVH is its sister station on the island of Kauai in the Hawaiian Islands. Both stations operate continuously on precisely 5, 10, 15, and 20 MHz. These stations are called "standard time and frequency" stations, because their operating frequencies are so accurate they can be used as laboratory-grade frequency standards. They also both broadcast the UTC time (as determined by an atomic clock) each minute with "ticks" each second between the minutes. The time on WWV is announced by a man, and on WWVH by a woman. The two stations are so perfectly synchronized in frequency and time signals that it's not unusual for listeners in the western half of the United States to hear both simultaneously without realizing it. To tell them apart, listen carefully for the announcement of the time just before each minute. The announcement on WWVH begins before the one on WWV, and its possible to hear the WWVH female voice "under" the WWV second ticks prior to the start of the WWV announcement. WWV and WWVH also broadcast regular voice announcements of current propagation conditions, including warnings about geomagnetic "storms," at regular intervals.

Now do you see why I said the ability to receive the HF ham and shortwave bands is an important part of any ham radio station?

Novice

The reduced Novice privileges simplify things a bit. If I had room to erect some sort of resonant antenna for 10 meters, such as a vertical or dipole, then I would have to get active on that band. If I could erect an antenna that covers 10 meters and other bands (like 40 and 15 meters), so much the better. But 10 meters alone would be enough to keep me happy and busy for a long time. I could operate SSB phone, packet, RTTY, and CW, and during years of high sunspot activity (like 1988 to 1991), it's possible to contact over 100 countries in a single weekend of a major DX contest. Even in years of low sunspot activity there's plenty of sporadic-E, and there's reliable coverage over a radius of 30 to 50 miles or so even when there's no skip.

It's possible to get started on 10 meters for about the cost of a 2-meter FM hand-held unit. There are a few inexpensive SSB transceivers that operate only 10 meters available. These usually have an output power of 25 watts or so, which is enough for local communications and worldwide DX when the ionosphere cooperates. An ordinary CB antenna can be slightly shortened by cutting away at the elements and be made resonant on 10 meters. For less than $300, I could be ready to talk to the world when the sunspots are willing.

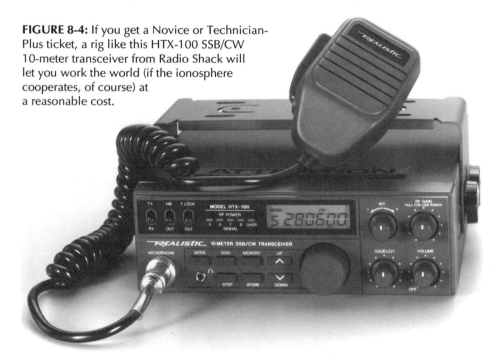

FIGURE 8-4: If you get a Novice or Technician-Plus ticket, a rig like this HTX-100 SSB/CW 10-meter transceiver from Radio Shack will let you work the world (if the ionosphere cooperates, of course) at a reasonable cost.

If my budget will let me afford something more expensive, I would get myself a SSB/CW transceiver covering at least 80 to 10 meters. Not only could I operate to the limits of my Novice HF privileges, but I would have the foundation of a ham station that could serve me well even after I obtained my Extra class license. Today's transceivers are really incredible. Most include a general coverage receiver covering from about 100 kHz to 30 MHz as part of the receiving circuitry, two VFOs (so you can transmit and receive on different frequencies), memories to store frequencies so the transceiver can be tuned to them by pressing a button, speech processing, RF wattmeter, and SWR meter in a single package. Many also offer FM operation on 10 meters (yep, there are some FM repeaters up around 29.6 MHz). Some are surprisingly inexpensive for all they offer—a transceiver offering all this can be bought for less than most laptop personal computers. And these transceivers can also operate from a 12 V battery, so you can install them in your car or operate portable using an auto battery.

However, what would I do if I lived in an apartment and couldn't operate HF? Novices are allowed FM phone privileges at 222 and 1240 MHz, including being able to work through repeaters. However, the usefulness of these privileges is limited. The activity on 1240 MHz is very low; large areas of the country (including entire states) have no repeaters for this band and even in a densely populated area (such as Los Angeles or New York) you might have only a handful of people to talk to. The situation is better on 222 MHz, but there are still shortcomings. This band is similar to 2-meters in its coverage, and there are numerous repeaters across the country. Unfortunately, the band is so similar to 2-meters that many non-Novice hams have never bothered to get active on it. I've used a 222 MHz FM hand-held to listen and check into repeaters in such areas as New York, Los Angeles, San Francisco, and Boston, and take it from me—there isn't a heck of a lot happening on that band. But if I couldn't get on HF with my Novice license, getting on 222 MHz FM would definitely be my second choice.

My first choice? I'd upgrade to a Technician class license and get on 2 meters as fast as I could!

Technician–Plus

Okay, now I've passed the Novice and Technician written exams and the 5 WPM code test. All the options outlined in the previous paragraphs are now open to me. So what would I do?

I'd still spring for a 2-meter hand-held as my first unit. That's one item I wouldn't outgrow and would get me yakking with the other folks ASAP. I'd also probably get a TNC for packet next if I had a PC, and then I'd add. . . .

No, not a 2-meter SSB/FM/CW transceiver. I'd add a 80 to 10 meter SSB/CW transceiver, particularly if it had FM operation on 10 meters. I'd especially do this if I could operate HF from my home, and I'd probably do it even if I lived somewhere (like an apartment) where I couldn't operate HF.

My reason for this is something we briefly mentioned in the last chapter. It's a transverter. This takes the output of a transceiver on a band like 10 meters and converts it to two meters or 70 cm. The reverse happens on receive.

Now transverters aren't exactly household items. They tend to be sold either by small specialty dealers or on the used equipment market (they were more popular years ago before the rise in popularity of FM at VHF/UHF frequencies). Or you can build your own or pay someone to do it for you (hint: this route isn't easy or cheap). But if you can get your hands on one for 2 meters, then a transverter being fed by a transceiver operating on 10 meters makes a great way to get active on that band, especially for SSB work.

FIGURE 8-5: If you're going to make 2 meters your main operating band, a transceiver like the Icom IC-275A is a good investment. It gives full SSB and CW operation as well as FM. It also has features that make it easier to use with a TNC for packet operation.

But let's suppose that I decided to go with a 2-meter SSB/FM/CW transceiver as my next purchase, perhaps because I couldn't find a transverter. Even if I couldn't operate HF from where I lived, I'd still get that HF transceiver as the third item in my shack. For one thing, a lot of ham satellites have an output ("downlink") on 10 meters with an input (uplink) on 2 meters. A HF transceiver would give me the receiving end to go with the 2-meter transmitting capability. The HF transceiver would almost certainly have a general coverage receiver, so I could keep in touch with WWV. And heck, I could probably get on 10 meters using low power (10 watts or less) by attaching a mobile CB antenna to a baking sheet, positioning it just outside my window, tuning it with an antenna tuner, and then removing the entire mess when I'm not operating. (Dropping the hypothetical again, I've actually operated 10 meters from the 36th floor of a high-rise condominium using that setup!) Finally, odds are I won't be living in a place with antenna restrictions forever. Sooner or later I'll be able to put up a decent HF antenna, and I'll want a rig that's ready to go when I am.

Anyway, that's what I'd do if I were in your shoes.

Ham Radio on the Cheap

I remember when I was a young ham (yeah, it was a long time ago!) and how I got ill when I looked at the price tags of new ham gear. Fortunately, hams are always upgrading and changing the equipment in their shacks, so there's always a large amount of used gear available which can get you on the air for a lot less than new stuff. But where do you find it?

Before trying to find used equipment, you need to get familiar with what's good and what's bad in older equipment. Back issues of magazines such as *QST* (and most larger libraries have several years of *QST*) have ads and equipment reviews that will let you get familiar with the rigs available in the past and their features. It also helps a lot to get to know some more experienced hams and ask their opinions. Finally, the popularity of an item of gear can be a good (though sometimes misleading) indication of whether it's good. If there are a lot of a certain type of rig available, that means quite a few were sold. And it probably wouldn't have sold so many if it hadn't been a good piece of equipment.

One great place is at ham equipment dealers. Many of these accept trade-ins of used equipment, and several maintain extensive selections of used equipment. If your local dealer doesn't have what you're looking for, you might write to one of the advertisers in the ham magazines that mentions they have a used equipment list available. Buying through a dealer is more expensive, but there are some advantages like being able to return the item within a certain period (such as 30 days) if there's a problem. Most dealers also check out their used gear and make any necessary repairs before shipping it to you. Even if your local ham dealers don't sell used gear themselves, many have bulletin boards where local hams put up notices of stuff they are selling.

Another good place to find used gear is at a hamfest and their "flea markets" where individuals buy and sell equipment. Here you have the advantage of being able to inspect and often use the equipment before you buy. If you're a good negotiator, you can also equip your station for only a fraction of the cost of new gear. However, the old *caveat emptor* cliche really applies here—if you buy something and find a problem when you get home, you're generally stuck. It's a good idea to have a more experienced ham along with you when you do the rounds of your first few hamfests. (Don't forget to do the same for another new ham in a few years.) And remember this simple rule: *try before you buy!*

There are several biweekly and monthly "trader sheet" publications full of classified ads from hams offering used items for sale or seeking certain items of discontinued equipment. These are a bit like a flea market by mail, although you aren't able to physically inspect the items before buying. If you go this route, insist on being able to return any item that doesn't meet the buyer's description and keep careful records of any correspondence, telephone calls, checks, or any other items pertaining to the purchase.

What to buy on the used ham equipment market could be a book unto itself. However, stay away from vacuum tube gear and stick with solid-state gear. Some vacuum tube gear can offer a lot of performance and fun (I'm still an aficionado of the stuff), but it has too many pitfalls (such as tubes that are no longer available) for newcomers. Good solid-state HF or VHF transceivers from 15 years ago to the present are usually safe choices. With some research in back issues of ham magazines, some looking and effort, and a little hard bargaining you can get on the air for a lot less than you might think!

How to Be a Smooth Operator

You want to be a ham radio *operator*. An operator is one who operates—meaning, as the dictionary says, to perform a function or task so as to bring about a desired effect. Operating a ham radio station is much more than turning on your equipment and letting it rip. How you operate your station is a lot more important to really enjoying ham radio than the exact items of gear you have!

If you don't remember anything else in this chapter, please remember this: *always listen before you transmit*. This goes whether you're operating on 160 meters or 70 cm, whether you're on CW or a FM repeater, or whether you're chasing a rare DX station or just want to join a conversation in progress on a repeater. Nothing will make or break your reputation among other hams quicker than whether or not you listen to a frequency before plopping a signal down on it.

One reason is simple courtesy. I can't tell you how infuriating it is to be in the middle of a pleasant QSO with an old friend—or one I just met—and then have some idiot interrupt it by calling CQ on the same frequency. This isn't too smart a way to make a QSO. The people already on a frequency aren't going to move aside and let an interloper use it, nor is anyone else listening going to respond to the CQ call.

You also miss a lot when you don't listen, particularly if you're into activities like contesting and DXing. That rare country or station you're trying to work might be lurking in some out of the way spot on a band. If you don't listen around carefully, you'll never know it's there.

Listening a bit before transmitting lets you get the lay of the land. Ever been talking to a group of friends when someone comes up to your group—maybe it's even someone you all know—and this person starts yakking away on some subject completely different from the one you and your friends were talking about? It's irritating when it happens in person, and it's irritating when it happens over the air. A little listening before you transmit, even if it's before joining a group of friends you hear, can help you have something to say that other people will be interested in hearing. (And that applies whether you "say" it in CW, RTTY, packet, or phone!)

Finally, pausing for a second or two before you reply to another station you're in contact with can be a good idea, especially if you're communicating through repeaters. That brief pause can let another station join your conversation, such as a friend who'd like to say hello or a station with an emergency that needs your assistance.

So take a couple of seconds to listen before you press the button on your microphone or press the keys on your RTTY keyboard. It's what separates the good operators from the dopes. Hams have a term for other hams who don't listen like they should—*alligators*. After all, one of the best descriptions of an alligator is something that's all mouth and no ears!

A lot of new hams (and too many more experienced ones) fall into a trap on the HF bands. They listen for a few seconds, hear nothing, and let out a CQ call. Instead of the contact they want, they are angrily told that a QSO is in progress on the frequency. This can happen because propagation at HF often lets you hear only one side of a QSO. The solution is to quickly ask "is this frequency in use?" on phone or "QRL?" on CW or RTTY before transmitting on what seems to be a clear frequency. If you get no reply after a few seconds, then you can safely go ahead and call CQ.

No One "Owns" a Frequency

A lot of ham radio operations, such as nets, take place at certain times on frequencies agreed upon in advance. That means that eventually most hams are placed in a situation where they are busily chatting away on a certain frequency when a station interrupts the QSO and informs them that in two minutes the East Central States Mobile Sideband Traffic Net is going to start on that frequency.

What to do? What do I do?

Let's make one thing clear: *no one owns a frequency in any ham band.* They're shared by all hams. The East Central States Mobile Sideband Traffic Net has no more right to that frequency than I do—and no less right either. The only exception is that *bona fide* emergency communications always has priority over all other communications.

Courtesy, cooperation, and common sense are the keys here.

Keep in mind some fundamental facts of human nature. If you ask someone to please do you a favor, most people will try to accommodate you. If you give an order—and you're not a parent, teacher, cop, or other authority figure with the clout to back up that order—people will defy and thwart you. If you politely ask someone to QSY or let you use the frequency for a moment, most hams will try to help you out. But give someone an order to QSY, and you've got problems. I've heard a couple of stations "ordered" off a frequency by a particularly obnoxious "net control" station refuse to budge and instead switch on their linear amplifiers and hold on to a frequency by brute signal strength!

Another thing to keep in mind is that there's a bit of "finders keepers" involved in frequency use. If two stations are in the middle of a QSO, no one else can deliberately operate in a manner so as to cause QRM on that frequency. You'd be surprised at how many nets just set up shop on a given frequency regardless of whether it's in use. But as far as the FCC is concerned, those nets and their stations are deliberately QRMing other stations and are asking to be cited for a violation of FCC rules and possibly fined.

If you're asked to move off a frequency because a net is about to start, and can do so without disrupting the QSO, it's polite to do so. It's also polite for the net to give you and the other station you're in contact with enough time to find another frequency. There are so many nets in operation that it's impossible to keep track of all of them, but do try to be familiar with the times and frequencies of the major ones and avoid operating on their frequencies near the starting time. And if you're part of a net, remember to be patient and courteous if someone is on "your" frequency when it's time for the net to start. It almost certainly wasn't done deliberately or as an act of aggression—and the stations already on the frequency have as much right to it as your net.

Try the golden rule. Treat your fellow hams the way you would like them to treat you.

When you do call CQ, use the procedure outlined in Chapter 2 and try shorter calls more often instead of lengthy, droning calls. Try calling CQ on phone, CW, or RTTY for no longer than 30 seconds or so at a time and then listen for a few seconds. If you don't hear anything, repeat the process and call CQ again for about 30 seconds. Don't think that a marathon CQ call will generate more replies or even a reply at all. When I tune around and run across one of those endless CQs, I figure that person is more interested in talking to themselves than to anyone else!

You might also want to take a tough, objective look at the way you talk, perhaps by making a tape of yourself and listening later. I'm not saying that you need an accentless, neutral voice, but you do need to be understood by other people. This doesn't mean you have to take voice lessons, but it does mean you have to get into the habit of speaking clearly. Be careful of speaking too rapidly or slurring your words, especially when giving your call sign, name, or other important data. Just listen on the HF bands, particularly in a contest period, at how often people are asked to repeat their call signs—and how those people speak when giving their calls! It's often hard to tell if they said "C," "V," "B," or "D," or if they said their name was "Larry," "Barry," "Jerry," or "Harry." So get that jaw working and enunciate when you talk! And remember to use the phonetic alphabet when conditions get rough or if the person you're talking with doesn't have a good command of English.

Get familiar with the operating procedures and conventions for the bands and modes you want to use, especially any applicable band plans (we'll cover them next). Different activities, such as DXing and contesting, have their own set of practices which constitute "good operating." For example, lengthy CQ calls on a repeater are out of place; a simple "this is AA6FW monitoring" or "this is AA6FW, anyone around?" is all that's needed to initiate a contact. You can read about such procedures in books like *The ARRL Operating Manual* and see how they're implemented in the real world by listening to the bands. (Here's some advance warning: what you read should be done and what you hear being done are often in conflict.)

There's no big secret to being a good ham radio operator. Just speak clearly and distinctly. Listen before you transmit. Listen to what the other hams are saying. Be familiar with basic operating procedures for the band and mode you're using. And, above all, keep in mind that we're all in this together and have to cooperate with each other. Basic decency, courtesy, and empathy for others will go a long way toward making ham radio more enjoyable for everyone.

Should Base Stations Talk Over Repeaters?

If you read much about what's supposed to be proper operating procedure on FM and repeaters, you'll see something like this:

"The purpose of a repeater is to extend the coverage of hand-held and mobile stations. Base stations should avoid using repeaters and communicate via simplex instead."

Some variation of that statement is a staple of many articles and book chapters on correct FM operating procedure. It's been repeated so often that many people take it as gospel.

Well. . . . nuts!!! I disagree. Rather strongly, in fact.

As a practical matter, this reasoning—if widely followed, which it is thankfully not—would mean it would be difficult, if not impossible, for a hand-held or mobile unit to ever contact a base station on VHF/UHF FM. After all, why would a base station ever monitor a repeater if they aren't supposed to communicate through it? That being so, then a good bit of the value of FM and repeaters is reduced. Those base stations really help in emergencies, since they can readily make telephone calls, summon assistance, get things, and do a lot of other useful tasks that mobile and hand-held stations can't. Even if it's not an emergency, why should I be restricted to talking to other hand-held and mobile stations over a repeater? What if my buddy is at home instead of in his car or walking around with a hand-held unit? Don't people at FM base stations have something useful or interesting to say at times?

One reason given to keep base stations off repeaters is that the 2-meter band is crowded and repeaters are too busy to handle base station communications that could go via simplex. But listen sometime to the repeaters in your area. One or two may be busy, but most are quiet. . . . . very quiet, in fact. Most repeaters could support a lot more activity than they do!

FM repeaters are great because they level the playing field. My hand-held or mobile unit can produce a signal just as strong as a base station on the output. And all stations within the repeater's prime coverage area are part of that level playing field.

The bias against base stations operating through repeaters is an artifact of the early days of FM and repeaters. I agree that if two base stations just want to talk to each other and nobody else, or if they're feeling especially long-winded, they should find a vacant simplex channel and yak away there. But why should they otherwise avoid repeaters if they operate according to established procedures?

Remember: just because people keeping saying something for years and years doesn't mean it's true. It might mean they've never stopped to think about it carefully.

Band Plans

In addition to the band allocations set up by the FCC for different modes and license classes, hams themselves have agreed to some de facto allocations of certain frequencies for different purposes. These "allocations" are unofficial and compliance with them is strictly voluntary. However, everyone wins if these *band plans* are followed, since certain legal activities aren't too compatible with other legal activities. Band plans also save time and headaches; you know in what frequency ranges you'll find other stations interested in what you're interested in. Here are some of the frequency ranges and frequencies on the HF bands you should be aware of:

kHz	Usage
1800 to 1830	CW and RTTY only
1818	W1AW CW frequency

kHz	Usage
1830 to 1840	International contacts only using CW and RTTY
1840 to 1850	International contacts only using all modes
3581.5	W1AW CW frequency
3590	RTTY DX frequency
3610 to 3630	RTTY only
3625	W1AW RTTY frequency
3767	"Geritol Net," active winter nights; a net for Extra class amateurs attempting to work all states on this band
3790 to 3800	"DX window" in which DX stations transmit and U.S. stations don't so the DX can be heard
3845	Slow-scan television
3880	AM phone operation, often by hams using restored gear from the 1950s
3905	Century Cub Net for those attempting to work all states
3990	W1AW phone bulletin frequency
7040	RTTY DX frequency
7047.5	W1AW CW frequency
7080 to 7100	RTTY only
7095	W1AW RTTY frequency
7171	Slow-scan television
7290	W1AW phone bulletin frequency
10140 to 10150	RTTY only
14047.5	W1AW CW frequency
14070 to 14099	RTTY only
14095	W1AW RTTY frequency

kHz	Usage
14100	DX beacons around the world to indicate propagation conditions
14230	Slow-scan television
14290	W1AW phone bulletin frequency
14336	County Hunters Net for people attempting to contact all U.S. counties
18097.5	W1AW CW frequency
18100 to 18110	RTTY only
18102.5	W1AW RTTY frequency
18160	W1AW phone bulletin frequency
21067.5	W1AW CW frequency
21070 to 21100	RTTY only
21095	W1AW RTTY frequency
21340	Slow-scan television
21390	W1AW phone bulletin frequency
24920 to 24930	RTTY only
28067.5	W1AW CW frequency
28070 to 28150	RTTY, AMTOR, and packet only
28095	W1AW RTTY frequency
28190 to 28225	DX beacons around the world to indicate propagation conditions
28590	W1AW phone bulletin frequency
28680	Slow-scan television
29300 to 29510	Satellite downlinks
29520 to 29580	FM repeater inputs
29600	FM simplex
29620 to 29680	FM repeater outputs

The band plans for 50 MHz and higher are really complex and vary in different parts of the country. A recent edition of

The *ARRL Repeater Directory* will contain all the latest ones, but here are some highlights you should be aware of:

MHz	Usage
50.06 to 50.08	Propagation beacons
50.11	DX calling frequency
50.2	National calling frequency
50.4	AM calling frequency
50.6 to 51	Experimental and special modes
51 to 51.1	"DX window" for Pacific stations; stations located in the Pacific and Asia transmit on this frequency while those in United States don't
51.1 to 52	FM simplex
52 to 52.05	"DX window" for Pacific stations
52.525	National FM simplex frequency
144 to 144.05	Moonbounce using CW
144.1 to 144.2	Weak signal SSB
144.2	SSB calling frequency
144.2 to 144.3	SSB operation
144.275 to 144.3	Propagation beacons
144.3 to 144.5	Satellites
145.5 to 145.8	Experimental and miscellaneous modes
145.8 to 146	Satellites
146.52	National FM simplex frequency
420 to 432	Television
432 to 432.07	Moonbounce
432.1	SSB calling frequency
432.175 to 432.175	Satellite uplinks
432.3 to 432.4	Propagation beacons
435 to 438	Satellites
446	National FM simplex frequency

Again, these band plans aren't mandatory. But they do help everyone have some space for specialized operations and keeps QRM between activities to a minimum. If you ask me (and even if you didn't, I'm writing this book so I'll answer anyway), only a real jerk would operate SSB on 146.52 MHz or call CQ just before W1AW is scheduled to start a code practice transmission. A little cooperation goes a long way!

Strange Ham Lingo—The Final Episode. . . .

There's no end to the number of unusual terms used in ham radio, but we are at the end of this book so this is our final installment of this feature. Always remember this important rule: if a fellow ham uses a word or expression you don't understand, *act like you do!*

beacon: a station that automatically transmits its call sign at specified intervals, operated to let other hams know if the band is open to a certain point.

bulletin: one of the few one-way ham transmissions permitted by the FCC; it must consist of material and information solely of interest to hams.

calling frequency: a frequency that most stations active on a given band monitor and listen to for CQ calls, etc. Once contact is established with another station, both QSY to another frequency and leave the calling frequency open for others. This is a great way to concentrate activity on "wide open" bands like six meters, etc.

DXpedition: an organized effort to put a rare DX country on the air by hams who travel to that country just for that purpose.

Elmer: a ham who generously donates his or her time to helping new hams get their license and start on the right foot.

EME: communicating by bouncing a signal off the moon and back to Earth, or "Earth–moon-Earth."

exchange: in a contact, passing the necessary information between stations, such as location and a signal report, for the contact to count for contest points.

gallon: the full legal power limit.

homebrew: home-built ham equipment.

junkbox: where hams keep spare parts, miscellaneous hardware, random items. . . . in other words, their junk.

lid: a crummy operator; that is, one who doesn't check to see if the frequency is in use, calls CQ too long, is rude, has a lousy (bad sounding and distorted, not weak) signal, and generally makes a complete pest of him or herself.

moonbounce: same thing as EME above.

net control: the ham operator who is the "master of ceremonies" for a net, including calling the net to order, recognizing stations checking into the net, and directing its operations.

pileup: when several stations are calling the same DX station and the result is heavy QRM to each other.

ragchew: to talk at length about any subject at all.

real ham: a ham who's always looking to try something new, uses correct operating procedures, helps newcomers, keeps a sense of humor and courtesy, tries to keep up with current technology, remembers that ham radio is only a hobby, still has that sense of wonder about it all, and no matter how long he or she has been on the air still gets a rush when they hear somebody distant calling.

sidewinder: old term for a SSB station. (Yeah, I can't figure this one out either.)

split frequency: transmitting on one frequency and listening on another, often done when chasing DX.

Ten-ten number: Ten-ten International is a ham club devoted to increasing activity on ten meters. They sponsor awards and contests based upon exchanging membership numbers, or Ten-ten numbers, with other hams. If you operate for any length of time on 10 meters, you'll be asked for your Ten-ten number. Better get one for self-defense.

tie the ribbons: to end a QSO. "Well, Bob, it's time for me to tie the ribbons on this one. . . "

window: a frequency range set aside for DX stations to transmit in while American stations do not, done to avoid QRM to the DX. This is an example of split frequency operation.

Awards

I was the sort of kid who always was sending off for a Captain Midnight decoder ring or membership in the Rin-Tin-Tin fan club. Maybe that's why I enjoy earning different awards for ham radio operation sponsored by organizations like the ARRL and ham radio magazines.

Major awards require you to contact and receive QSL cards from stations located in different countries, states, geographic zones, etc. Most of these awards can be "customized" by working the required number of stations on a certain band (all 20 meters, for example) or mode (all SSB or phone), so you can make an award even more challenging. The award itself comes as a handsome certificate suitable for framing; a few of the more difficult and prestigious awards issue wall plaques to the recipients.

One award you can shoot for regardless of your license class is the *Worked All States* (WAS) issued by ARRL. As its name says, you have to contact and QSL another ham station in each of the 50 states. This popular award is one you can earn in any number of ways, such as all RTTY or satellite contacts, all 40 meters or 6 meters, or even all 40 meters RTTY. There's also a special five-band WAS for contacting each state on 80, 40, 20, 15, and 10 meters.

Another ARRL award is perhaps the most sought-after award in ham radio—the DX *Century Club* (DXCC) issued for contacting 100 or more different countries. As we've mentioned before, hams define a "country" differently from normal people; states separate from the mainland (Alaska and Hawaii), U.S. possessions (Puerto Rico, the Virgin Islands, and Guam), and isolated mid-ocean bits of rock owned by countries such as France and Norway are all considered separate countries. Maybe that's not too logical, but the aim is to give everybody as many different targets to shoot for as possible. Currently, there are over 300 radio countries recognized by the ARRL for the DXCC award. DXCC can be endorsed for making all contacts in a certain mode (such as phone or RTTY), band (such as 160 meters or 6 meters), or by satellites. There's also a five-band DXCC for contacting at least 100 countries on 80/40/20/15/10 meters.

FIGURE 8-6: Many hams frame and proudly display their DXCC certificate on the walls of their shack. Some even go so far as to use it as an illustration in books they write.

A more difficult DX award is the *Worked All Zones* (WAZ) sponsored by CQ Magazine. CQ has divided the world into 40 roughly equal-sized geographic zones, and WAZ requires you to contact and QSL a station in each zone. This is a tougher feat than DXCC, since WAZ requires you to work stations in the Middle East, sub-Saharan Africa, the Indian Ocean, southeast Asia, and similar tough spots that you can ignore for DXCC. There is also a five-band WAZ award, which requires you to earn WAZ on 80, 40, 20, 15, and 10 meters. This latter award is generally considered to be the toughest challenge in DXing; quite a few people have earned five-band DXCC but have a long way to go for five-band WAZ.

Maybe the easiest DX award for newcomers to ham radio is the *Worked All Continents* (WAC), sponsored by the International Amateur Radio Union. In the United States, it's available through the ARRL. It requires you to work and QSL at least one station in North America, South America, Europe, Africa, Asia, Australia, and Oceania (Antarctica is not counted due to the extreme rarity of ham radio operation from there). You can also get this award endorsed in multiple ways and in a five-band version.

Other awards are available for contacting 500 or more U.S. counties, at least 300 or more call sign prefixes, or for assorted numbers of European countries, Canadian provinces, Japanese prefectures, Russian oblasts (similar to U.S. counties), and almost any geographic or political entities that can be counted. Most hams stop after earning a couple of major awards (like DXCC and WAS) while others can't stop until their shack walls are covered with wallpaper. But be careful—if the certificate chasing bug bites hard, it clamps down and doesn't let go!

Contesting

Another way to indulge your competitive instincts is through on-the-air contests. These are usually over a weekend (such as 0000 UTC Saturday to 0000 UTC Monday) and require you to work as many different stations, countries, zones, states, etc., as you can within that period. Your final score is usually the number of different states worked multiplied by the number of different countries, states, zones, etc., contacted. You also have to exchange some information with the other station for the contact to be valid for contest credit. A signal report (invariably "5 by 9" regardless of the actual signal) is almost always required along with your zone, state, a consecutive QSO number, etc.

Internationally, the biggest contests are the Worldwide DX (WWDX) contests sponsored by *CQ* Magazine. The phone contest is held in October, followed by a CW contest in November. They also recently added a RTTY contest in September. These contests attract thousands of contestants from around the world and stir up a great deal of activity. It's not too unusual for a well-equipped station to contact well over 100 countries in a single contest weekend. Another active DX contest is sponsored by the ARRL. The CW section of the ARRL contest is held in February followed by the phone section in March.

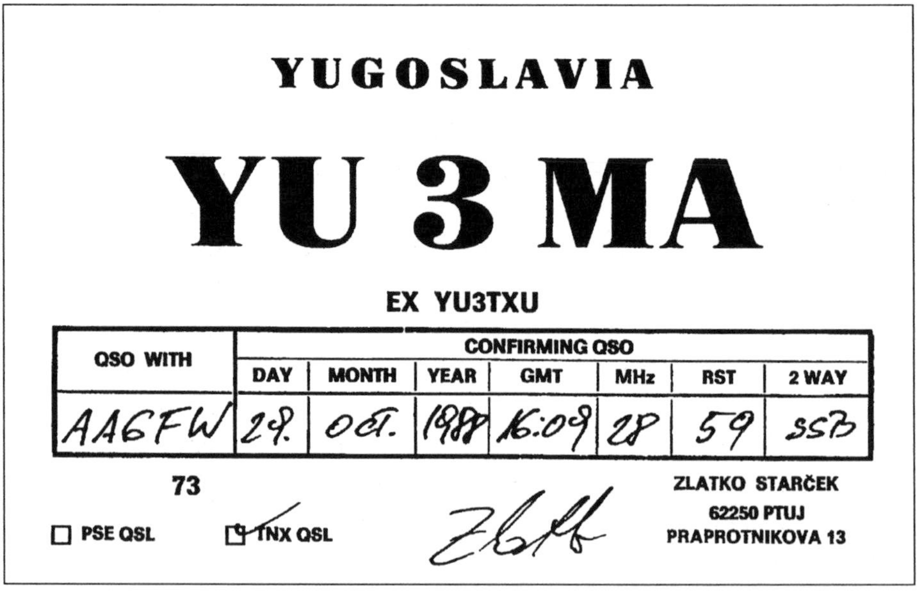

YUGOSLAVIA

YU 3 MA

EX YU3TXU

QSO WITH	CONFIRMING QSO						
	DAY	MONTH	YEAR	GMT	MHz	RST	2 WAY
AA6FW	*29.*	*Oct.*	*1988*	*16:09*	*28*	*59*	*SSB*

73

☐ PSE QSL   ☑ TNX QSL

ZLATKO STARČEK
62250 PTUJ
PRAPROTNIKOVA 13

FIGURE 8-7: This QSL card confirms a contact that took place during *CQ* Magazine's Worldwide DX Contest back in 1988. Contests like this are great opportunities to work a lot of DX in a hurry.

The ARRL sponsors a popular domestic contest known as the *ARRL Sweepstakes*. There are separate weekends for phone and CW each November. The ARRL has divided the United States and Canada into different sections according to population (for example, California has been divided into nine sections) and the object of Sweepstakes is to contact as many different sections and stations as possible. The real goal of most participants, however, is a "clean sweep"—contacting at least one station in each ARRL section. The ARRL also sponsors domestic VHF and UHF contests, a RTTY contest, and a "Novice roundup" open to Novices and Technicians only.

Many other organizations sponsor international and domestic contacts. There are contests where the rest of the world tries to work as many Japanese or European stations as possible as well as various state "QSO parties," where the object is to work as many different stations from a given state as possible. Full details and rules for upcoming contests can be found in any issue of CQ or QST.

Even if you're not into competing to win, contests offer a terrific way to quickly build up your total of different countries or states worked. There are seldom pileups in a contest, and QSOs tend to be quick so everyone has a chance to work a rare country. Moreover, many stations are just as eager to contact you (for the QSO points) as you are to contact them. While I'm no threat to ever win a contest, I do enjoy keeping my operating skills sharp in a contest environment and handing out QSOs and points to those who are in the contest. Contests aren't for everyone—for instance, it's hard to find a clear frequency for a decent ragchew on 20 meters during the CQ WWDX contest—but they might be for you.

You've Finished This Book— Now Read Another

You're near the end of this book, but not at the end of your learning more about ham radio. As mentioned back in Chapter 1, you need to equip yourself with a good study guide or two to help you prepare for the latest exam for the license you want. There's always something new to learn and try. The best way to learn new things and how to do them is through a good book. Almost all of the books we'll quickly look at in this section are available from your local ham radio dealer or mail order dealers advertising in the ham magazines.

The very first new addition you should make to your ham library is the latest copy of the ARRL *Handbook for Radio Amateurs*. The *Handbook* is quite simply the best bargain around, answering just about any question you might have on any subject in ham radio. My only criticism is that it is too top-heavy with construction projects; it sometimes seems from the *Handbook* that our hobby is amateur radio building instead of amateur radio operating. Fact is, most hams today don't build anything other than simple one or two integrated circuit projects or antennas, yet the *Handbook* always has several humongous projects requiring one to be a skilled electronics craftsperson/sheet metal worker. Despite this, this book is essential and a terrific value.

Another great ARRL book is *The ARRL Operating Manual*, which should be required reading for all ham radio operators. Everything from DXing to contesting to emergency communications to traffic handling is covered in a lively, cogent manner. If you're like most hams, you'll also want to add a copy of *The ARRL Antenna Book* to your library. Many of the designs in the book are well within the capabilities of any ten-thumbed ham to whip together from some wire, aluminum tubing, coaxial cable, simple hardware, and determination.

How do you find the addresses of U.S. and foreign amateurs for QSLs? *The Radio Amateur Callbook* is published annually in two editions: one for North America (U.S., Canada, Bermuda, Mexico, U.S. Possessions, etc.) and an international edition. New editions are published each December.

Finally, this book hasn't discussed many of the techniques for working DX. That's because Bob Lochner has written a superlative book called *The Complete DX'er*. Bob gives you all the techniques for working rare DX, but what really sets this book apart is Bob's way of conveying the adventure and excitement of DXing. Even if you don't get that interested in DX, you will enjoy reading this book!

Beware of Dinosaurs

I wish I didn't have to write this, but it's a subject I have to address. If you're going to go for a code-free Technician license, look out for the dinosaurs. You know what dinosaurs are, don't you? They were large, stupid beasts with small brains who couldn't adapt to changed circumstances and perished. We unfortunately have a few running loose on the ham bands who insist that ham radio is headed straight to hell because a code-free license is now available.

If you run across one of these types, don't try to argue with them or even attempt to rationally discuss the issue with them. These people have the mindset of a religious fanatic and are impervious to logic and common sense. They "know" that a CW test is all-important, and that it has the power to magically keep any troublemakers out of ham radio. (Yeah, it doesn't make sense to me either. . . .) But here's something funny: despite their belief in the importance of CW, if you look in the *Callbook* you'll see most of these types hold licenses requiring CW proficiency of 5 or 13 WPM. It always seemed to me that if they really feel the code was that important they would all have an Extra (after all, it requires a 20 WPM test!). Guess I was wrong.

If you get a code-free Tech ticket, just ignore such idiots. I'll handle them for you—I've got an Extra ticket and will challenge them to get one for themselves. (It's an old American tradition I'm trying to revive called *put up or shut up, put your money where your mouth is*, or something like that.) In the meantime, you can enjoy ham radio without having to deal with people who have a week-old salad for brains. After all, the future of ham radio belongs to the code-free newcomers, and one day we can all go to a museum and see the dinosaur bones.

73 for Now. . . .

It's time to tie the ribbons on this book-length QSO. It's been a long-winded one on my part, and now I'd like to hear from you. When (not *if*, but *when*) you get on the air, look around for AA6FW. I'm mainly on SSB and RTTY on HF, and I like to check in on local repeaters as I travel. Welcome to ham radio, and I hope it becomes as big a part of your life as it has mine!

About the Author

As a ham radio operator, Harry Helms holds the Extra Class license and has held his current call sign of AA6FW since 1988. He has previously held the calls KR2H, KA5M, AK4P, WA4EOX, and WN4EOX. Since moving to California, he has earned DXCC, WAC, WPX, and the CQ Magazine SSB DX Awards. Besides chasing DX on HF, Harry enjoys contesting, RTTY and packet, and ragchewing on 2 meters FM while traveling. Harry is also a well-known shortwave listener and is the author of the definitive guide to shortwave radio listening, *Shortwave Listening Guidebook*, also published by HighText. He is the author of over 200 articles for such magazines as *Popular Communications*, *Popular Electronics*, *Science PROBE!*, *Modern Electronics*, *Elementary Electronics*, *Science and Electronics*, *73*, and others. He has written over 20 other books on various topics in radio, electronics, and computing for such companies as McGraw-Hill, TAB, Howard Sams, and Prentice Hall. When not talking on his ham radio station, he can be found scuba diving, hiking, looking at the stars through a telescope, or continuing in his quest to climb all southern California peaks that are over 10,000 feet in elevation.

Over 750,000 Forrest Mims Fans Can't Be Wrong!

The Forrest Mims Engineer's Notebook
by Forrest M. Mims III

That's right, previous editions of *The Forrest Mims Engineer's Notebook* have sold over 750,000 copies worldwide. Continuing in that winning tradition, this updated edition is a goldmine of integrated circuit applications and ideas!

This isn't like any other book you've ever seen about IC applications. Forrest has carefully hand-drawn and hand-lettered the pages to recreate the "feel" of one of his actual laboratory notebooks. He has built and tested each circuit in this book. Forrest also includes numerous tips on parts substitutions, possible modifications, and circuit operation. The result is a practical, no-nonsense guide based on his years of intensive hands-on experience with IC circuits.

You'll find a wide array of proven circuit designs in this book, ranging from simple digital logic networks and amplifiers to rhythm pattern generators, tone decoders, temperature sensors, digital to analog converters, counters, bus transceivers, and many other useful circuit ideas. Each comes complete with IC pin numbers, values for other components, and supply voltages clearly indicated.

Over the years, *The Forrest Mims Engineer's Notebook* has been an essential reference for professional design engineers, educators, technicians, students, circuit hobbyists, or anyone else who needs concise, accurate information on different chip applications. This is one book that won't sit on the bookshelf—it will find a permanent place next to the logic probe, multimeter, and breadboard in your electronics lab!

The Whole World's Talking on Shortwave Radio—
and Here's How to Listen In!

Shortwave Listening Guidebook

by Harry Helms

As dramatic events rocked the world in the late 1980s and early 1990s,
shortwave radio listeners (SWLs) had front row seats to history. SWLs
were there in October of 1990 the moment West and East Germany
were reunited. In June of 1989, SWLs heard an announcer of Radio
Beijing's English language service courageously denounce the massacre
in Tien'anmen Square—an announcer that was never heard again on
the air. And in August 1991, SWLs around the world were glued to Radio
Moscow as hardliners made a final stand against democracy—and lost.

The world will continue to change, and shortwave will continue to
put listeners in the middle of the action. But how do you know which
shortwave radio you need? When and where do you listen?

Here's another winning book from the author of *All About Ham Radio*!
Shortwave Listening Guidebook tells you when, where, and how to listen to
the world on the shortwave radio bands. In non-technical language, you'll
find information on such topics as:

- selecting the shortwave radio that's right for you
- how reception conditions vary at different times of the day and year
- what the different controls on a shortwave radio do and how to use
  them
- profiles of major international broadcasting stations
- getting program schedules and other materials from shortwave
  broadcasters
- frequencies used by the United States armed forces, Air Force One,
  and ship-to-shore stations

You'll also find information on antennas. . . time signal stations. . .
"pirate" and clandestine radio broadcasters. . . and the hobby aspects of
shortwave listening. It's everything and anything a SWL needs to know!

Shortwave Listening Guidebook (ISBN # 1–878707–02–7) is available for $16.95
from shortwave and ham radio equipment dealers as well as technical book-
stores. If your local dealer doesn't have it, you can order direct from HighText
Publications, 7128 Miramar Road, Suite 15, San Diego, CA, 92121. Add $3.00
for shipping and handling ($4.00 for orders to Canada, $5.00 for airmail to the
rest of the world). U.S. funds only please. California residents please add sales tax.

Index

Technician class license, 30-31
Television communications, 71
Television interference, see Radio
    frequency interference
Terminal node controller (TNC),
    119-121
Test and measurement equipment:
    crystal calibrator, 237
    frequency counter, 237-238
    marker generator, 237
    multitester/multimeter, 238-240
    oscilloscopes, 96-97
    reflectometer, 228-229
    RF signal generator, 237
    wattmeters, 236-237
Test signals, 94
Third-party traffic:
    definition of, 74
    restrictions on, 75
Tinkering, 16
Transceivers, definition of, 27
Transistors, 206
Transmission lines, see Feedlines

Transmitters:
    CW, 210
    definition of, 27
    FM, 212
Triode, 206
Tuning up procedures, 231-232

Ultra high frequency (UHF) bands
    defined, 27

Very high frequency (VHF) bands
    defined, 27
Volunteer examiner coordinators
    (VECs), 32, 45
Voltage, defined, 180

Watt, defined, 189
Wavelength, defined, 25-26
WWV, 256
W1AW, 255-256

Yagi antenna, 218-219